SECONDARY CITIES IN DEVELOPING COUNTRIES

Volume 145, Sage Library of Social Research

RECENT VOLUMES IN
SAGE LIBRARY OF SOCIAL RESEARCH

SECONDARY CITIES IN DEVELOPING COUNTRIES

Policies for Diffusing Urbanization

Dennis A. Rondinelli

Volume 145
SAGE LIBRARY OF
SOCIAL RESEARCH

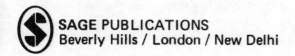

SAGE PUBLICATIONS
Beverly Hills / London / New Delhi

307,76
R77A

To Soonyoung, Linda, and Lisa

For information address:

SAGE Publications, Inc.
275 South Beverly Drive
Beverly Hills, California 90212

SAGE Publications India Pvt. Ltd.
C-236 Defence Colony
New Delhi 110 024, India

SAGE Publications Ltd
28 Banner Street
London EC1Y 8QE, England

Printed in the United States of America

Library of Congress Cataloging in Publication Data

Rondinelli, Dennis A.
    Secondary cities in developing countries.

    (Sage library of social research ; v. 145)
    Includes bibliographical references and index.
    1. Underdeveloped areas—Cities and towns. 2. Underdeveloped areas—Urbanization. 3. Underdeveloped areas—City planning. I. Title. II. Series.
HT153.R67   1983   307.7'63'091724   82-23165
ISBN 0-8039-1945-X
ISBN 0-8039-1946-8 (pbk.)

FIRST PRINTING

CONTENTS

PREFACE

A growing number of governments in developing countries have, over the past few years, been exploring ways of building the capacities of secondary cities to contribute to rural development and to a more diffuse pattern of urbanization. In response to this trend, the United States Agency for International Development (AID) commissioned a study of secondary cities; this book is a revised and reorganized version of that study. The objective was to review the literature on urbanization in developing countries and on the potential for middle-sized cities to contribute to national and regional development. The book draws on primary data, secondary literature, government documents, aid agency policy statements and project reports, and case histories of cities in developing countries. It brings together information on development policies, the dynamics and processes of secondary city growth, and the functions that these urban centers perform in national and regional development.

In conducting this study I received assistance from Eric Chetwynd, Jr., who helped with logistics and commented on various drafts, and from Marcus D. Ingle, who enthusiastically supported its initiation. A number of people read and commented on different versions of the manuscript, including Emily Baldwin, Ray Bromley, Sung-Bok Lee, Robert Mitchell, Randall W. Roeser, and D.R.F. Taylor, although none of them or the U.S. Agency for International Development should be held accountable for my conclusions, interpretations, or recommendations. I owe a special debt to Robert A. Hackenberg, from whose studies of Davao City in the Philippines I learned a great deal, for his comments and suggestions at two important stages in the book's evolution. Of course, I am also grateful for the patience of my family, to whom the book is dedicated.

This study draws on case histories of secondary cities that were written for much different purposes than my own. Their authors should not be held responsible for my interpretation of their analyses or for the context in which I placed the lessons derived from their work. Indeed, some of them may disagree with the arguments that I make using illustrations, examples, or data from their studies.

No one can do research on secondary cities in developing countries and not come away convinced that we still know relatively little about them and their potential for development. This study may raise as many questions as it seeks to answer, and that was in part its purpose. In the end, I had to be content to conclude, with Georg Simmel, that "nothing more can be attempted than to establish the beginning and the direction of an infinitely long road. The pretension of any systematic and definitive completeness would be, at least, a self illusion." Secondary cities offer a new frontier for research and policy analysis into which few scholars have ventured. But if this study does no more than intensify the debate over the role of secondary cities in regional and national development, it will have served a good purpose.

—Dennis A. Rondinelli
Syracuse, New York

CHAPTER 1

WHY SECONDARY CITIES ARE CRITICAL FOR NATIONAL DEVELOPMENT

International aid agencies and governments in many developing countries have turned their attention away from cities
and the problems of rapid urbanization to pursue policies for
alleviating rural poverty. The neglect of cities has been in part
a reaction to the "urban bias" of foreign aid and national
development policies of the 1950s and 1960s, and to the
increasing poverty in rural areas of Third World nations
during the 1970s. It has been reinforced by evidence of the
growing disparities in living conditions and levels of development between urban and rural regions, and by the conviction
of many planners that some cities in developing countries are
growing to an unprecedented and dangerously large scale.
The emphasis on rural development has also been an inevitable reaction to the failure of capital-intensive, export-
oriented industrialization strategies in many developing
countries and to the concentration of wealth in large metropolitan centers. Many development theorists believe that
rural impoverishment results directly from the growth of the
largest cities and that migration to them can be slowed or
deflected by strengthening rural economies.[1]

Thus in some developing countries attempts have been
made to halt the growth of the largest metropolis and to
disperse economic activities, a strategy that has either been

ineffective or extremely slow to produce results.[2] In other countries, investments have been reallocated to lagging or depressed regions, a politically controversial strategy that many economists claim can retard national economic growth. Some governments have promoted new towns and urban "growth poles" by concentrating industrial investment in regional or provincial capitals and by offering financial incentives to firms locating in industrial estates far from the largest city. But few of these policies have been successful in countries lacking systems of secondary cities that could support economic activities requiring high population thresholds and extensive services and facilities. Thus in most countries growth-pole policies have done little to develop rural hinterlands, and in some they have exacerbated regional economic dualism.[3]

Attention is now being refocused on the growing challenges of urbanization and on the relationships between urban and rural development. It has become more apparent in recent years that rural development goals, no matter how carefully conceived, cannot be achieved in isolation from the cities or entirely through "bottom-up" stimuli. Economic growth with social equity requires both accelerated agricultural development and expansion of urban industry and commerce. Greater attention must be given to diversifying the economies of small towns and middle-sized cities in promoting a more balanced distribution of income. The ties between urban and rural economies that have been crucial in promoting widespread development in richer countries are likely to become more important as rapid urbanization continues in the developing world. These linkages are crucial because the major markets for agricultural surpluses are in urban centers; most agricultural inputs come from organizations in cities; workers seek employment in towns as rising agricultural productivity frees rural labor; and many of the social, health, educational, and other services that satisfy basic human needs in rural areas are distributed from urban centers.[4]

Moreover, the poor in developing countries also live in cities. Thus the United States Agency for International Development, which has a mandate to help alleviate poverty in the developing world, has recognized the need to "deal with some of the critical problems of the urban poor which constrain development and to understand better the process of urbanization as it relates to development."[5] The World Bank has conceded that although rural development is essential to promoting growth with equity, it is "unlikely to cause significant reduction in the rapidity of growth in urban population or the problems of larger cities over the next two decades."[6] International agencies are now giving more attention to the complex linkages between urban and rural economies and to what World Bank analysts call "the important opportunities for increases in productivity and incomes and for a reduction in the incidence of poverty" that urbanization offers.[7] Policy analysts in international agencies now recognize the potentially positive role that a more balanced pattern of urbanization might play in promoting equitable development. "Urban and rural development have to go hand in hand, particularly if it is kept in mind that the definition of urban includes the whole hierarchy of cities from the metropolis through secondary cities to the 'rural-urban interface' of small cities or rural growth centers," Lubell argues. He notes that "cities of different size ranges pose different sorts of problems and serve different functions in the national economy. They are all relevant in a strategy for balanced growth."[8]

Thus as strategies are reformulated, they are also being refocused on a part of the urban system that had previously been ignored—secondary cities that are large enough to perform important economic and social functions for their own populations and those in surrounding areas, but that are smaller than the largest metropolises to which most of the benefits of economic development assistance had previously flowed. Some international assistance officials now call for a long-range policy of strengthening networks of intermediate cities, which "would have the dual advantage of dispersing urban populations, perhaps thereby reducing the flow toward the primate cities and building up other centers of urban activity which can serve as regional catalysts and accelerate the geographical spread of development."[9] The World Bank is including infrastructure and managerial assistance projects for secondary cities in its lending portfolio as policies for secondary city development appear more frequently in national plans.

An underlying assumption of the emerging policies is that a system of functionally efficient intermediate cities linked to larger and smaller urban centers and to a network of rural service and market towns can make an important contribution to achieving widespread economic growth and an equitable distribution of its benefits in both capitalist and socialist societies. But relatively little is known about secondary cities or the roles they play in national and regional development. Scholarly research on urbanization in developing countries has focused almost exclusively on large metropolitan areas and primate cities. Little attempt has been made thus far to review and compare the few studies that have been done of individual secondary cities. Thus, in most developing countries, policies for strengthening intermediate cities have been ad hoc and largely reactive attempts to deal with specific problems of regional inequity or metropolitan congestion as they arise.

This study examines the characteristics of secondary cities in developing countries, the functions they perform, their strengths and weaknesses in the spatial systems and economies of developing countries, factors affecting the dynamics of their growth, and their potential for serving as catalysts for regional and national development.

This chapter outlines the rationale for secondary city development policies that are part of a national strategy for achieving a more balanced pattern of urbanization and economic development. In Chapter 2, the term "secondary cities" is defined, the demographic, social, and economic characteristics of these cities are described, and some of the reasons that they currently play a relatively weak role in regional and national development are explored.

Appropriate development strategies for secondary cities must be forged from a better understanding of their processes and dynamics of growth. Thus in Chapter 3 the factors that have historically accounted for population growth and economic diversification in secondary cities are described, and it is argued that the dynamic relationships among these factors—the process of synergism—are what contribute to their sustained development.

The social, economic, and physical functions that contemporary secondary cities play in regional and national development are identified and described in Chapter 4, as are their weaknesses and limitations in promoting geographically widespread development. Given these characteristics and functions, and strengths and weaknesses, a broad strategy for developing a system of secondary cities is outlined in Chapter 5. The effects of colonialism and distortions in the allocation and distribution of national investments on urban growth patterns are examined, and the factors that seem to distinguish developmental from "exploitational" cities are analyzed. From this analysis, three elements of a national strategy are identified and described: building the economic and administrative capacities of existing secondary cities; developing small towns and market centers to perform urban functions more effectively; and strengthening the linkages between secondary cities and their rural hinterlands to stimulate a pattern of deconcentrated urbanization that can invigorate rural economies. The limited spread effects of urban economic activities are noted in Chapter 5, and it is suggested that secondary city development strategy must go beyond creating a few urban "growth poles." Unless development policies seek to strengthen indigenous economic activities and to diversify them in appropriate ways, they will either have little impact or be perverse. Unless the economies of secondary cities are integrated with those of their regions, through appropriate linkages with smaller cities, towns, and rural villages, their growth may generate strong "backwash effects" that drain their hinterlands of resources. Suggestions for strengthening the economies of secondary cities and for building on functions that they are capable of performing are outlined in Chapter 6.

The arguments for developing secondary cities are inextricably related to debates over the most appropriate spatial structure for promoting economic growth with social equity. In part they evolved from a reaction against the contentions of many economists that the concentration of capital investment in the largest metropolitan areas is the most efficient strategy for promoting development and that high rates of output resulting from such a strategy will automatically reverse spatial polarization and spread the benefits of development through trickle-down effects. They also reflect a growing dissatisfaction with optimal city-size theories and with prescriptions that rank-size distributions of cities found in Western industrial countries should always be encouraged in developing countries. Moreover, they question the efficacy of theories calling for widespread distribution of national investments in rural areas to stimulate "bottom-up" development through agropolitan or "spatial closure" policies.

Advocates of more balanced urbanization contend that trickle-down processes have not been effective in most countries and cannot be made effective without a well-articulated and integrated system of settlements through which innovation and economic stimuli can be diffused. Policies for dispersing investments widely usually ignore the scarcity of resources in developing countries and fail to produce the critical mass needed to generate economic and social change. Through the development of secondary cities, an attempt is made to promote a more deconcentrated or diffuse pattern of urbanization that can prevent or reverse polarization and stimulate rural economies. It recognizes the need for national governments to commit their substantial resources to strengthening intermediate and small-scale cities through "top-down" planning, as well as the need for strategic investment in social and physical infrastructure in rural market towns and villages to facilitate "bottom-up" development.[10]

Deconcentrating Urbanization

Many economists have argued that in poor countries where capital is scarce the highest rates of return on investments are obtained in the largest metropolitan areas, and that therefore primate cities should be encouraged to grow. Wheaton and Shishido contend that if developing nations seek to maximize economic output then "the optimum population size for a metropolitan area peaks at just over 20 million," when per capita GNP reaches about $2000.[11] Mera has argued that the growth of large cities in developing countries should not be inhibited because the marginal costs to the nation are small compared to increasing marginal benefits.[12] The results of these and other studies have led economists to conclude that the vast sizes to which many metropolitan areas in the developing world are growing are not economically inefficient. The high returns to investments concentrated in large cities will stimulate the growth of GNP and, through spread and trickle-down effects, the benefits will accrue to the rural poor.

However, much of the research on optimal city size—that is, the population size at which marginal social benefits equal marginal social costs—has been inconclusive. Concentrated investment theories came under increasing attack during the 1970s as development policies turned away from the goal of maximizing economic growth and sought more moderate rates of economic expansion with socially equitable distribution of benefits. If one goal of equitable growth strategy is to provide public and social services to larger numbers of people, then much of the city-size research indicates that small and intermediate cities offer economies of scale for efficient investment in public utilities, infrastructure, and social services. Economies of scale for municipal services in Western industrial countries are realized in cities with populations of from 100,000 to 300,000; cities in the 100,000 to 200,000 population range can supply all but the most specialized social services efficiently.[13] Studies of India have concluded that the costs of infrastructure needed for industrialization decline as a fraction of output in the 20,000 to 1 million population range and that economies of scale for most services are reached in cities of about 130,000 people.[14]

The assumption that the benefits of investments concentrated in the largest metropolitan centers would trickle down and thereby reduce urban-rural and interregional disparities has also come into question. Experience in most developing countries indicates the contrary—that growth of large metropolitan centers has often produced "backwash effects," which drain rural hinterlands of their capital, labor, and raw materials. Hansen argues that

> the trickling down of modernization has not reached the poor, especially in rural areas, or else has yielded them no more than marginal benefits. . . . The optimistic view that economic growth would result in a convergence of regional per capita incomes has not been supported by evidence.[15]

And, he notes, "regional income convergence could still be consistent with little gain in, or even a lowering of, the real incomes of the poorest groups, with a worsening of real income disparities."[16]

The failure of development to spread is attributed, at least in part, to inadequately articulated and integrated settlement systems through which innovation and the benefits of urban economic growth must be diffused. In much of the developing world, impulses from the center cannot spread, but, as Berry observes, "growth and stagnation polarize; the economic system remains unarticulated."[17]

The diffusion of development stimuli from both the bottom up and the top down in Western industrial societies occurred through hierarchical systems of cities with distributions that formed a lognormal curve, conforming to what is known as the rank-size rule. In countries with rank-size distributions, the population size of any city is inversely proportional to its rank in the hierarchy: The second largest city is expected to be half, the fourth ranking city about one-fourth, and the hundredth ranking city one-hundredth the size of the largest city. In such a distribution there are a small number of very large cities, more numerous middle-sized cities, and large numbers of small cities and towns.[18]

The regularity of this distribution in so many countries and regions is best explained by central place theory. Brian Berry, who has contributed much to contemporary research on central place theory, points out that in economically advanced countries most consumers are also producers of goods and services. Production of a wide variety of goods and services requires division of labor and locational specialization, which both result from and are causes of locational disparities in production capability. Thus in any society there tends to be a high degree of differentiation in production capacity among cities, towns, and regions, while consumers usually demand a similar set of products and services. In order for goods that are produced in specialized locations to reach consumers in other places, there must be a well-articulated system of distribution and exchange; goods must be assembled through local collection points, dispersed through distribution points, and made available to consumers through local markets. These points form a system of settlements that are part of a complex web of production, distribution, and exchange.

Most consumers want to go to those locations that minimize their travel time and costs, and thus there is a need for many distribution points spread over the landscape providing frequently used, or "lower-order," goods such as basic necessities and daily consumption items. Other locations must provide a larger number and greater variety of goods and services, more durable and less frequently purchased products, for which people are willing to travel greater distances. Settlements providing these "higher-order" goods and services must be larger in population size or be accessible from greater distances to attract sufficient numbers of consumers to support the broad range of goods they offer. Settlements with sufficiently large populations to perform functions that serve consumers from outside their own boundaries are called "central places."

As the system of central places becomes more articulated, with exchange taking place among smaller centers offering lower-order goods and larger centers offering higher-order goods, the system becomes integrated. Lower-order centers are "nested" within the hinterlands or trade areas of larger centers. As Berry notes, "These levels of centers define a central place hierarchy in which there are distinct steps of centers providing distinct groups of goods and services to distinct market areas. The interdependent spatial patterns of centers of different levels, and the interlocking market areas of goods and services of related orders weld the hierarchy into a *central place system*."[19]

Through this central place hierarchy—which often results in a rank-size distribution of cities—the specialized goods and services of particular locations are made available to consumers throughout the country. People living in or near small towns have access to lower-order goods and services in local markets as well as to higher-order functions that have to be located in bigger cities because they require higher population thresholds. A deconcentrated, articulated, and integrated system of cities therefore provides potential access to markets for people living in any part of the country or region, and it constitutes a "balanced" pattern of urbanization. But many developing countries do not have efficient systems of central places. Some lack sufficient numbers of such places. Others have many but they are not well integrated or linked. Many countries have settlement systems in which the distribution of cities is extremely skewed and is dominated by a "primate city"—a metropolitan area that is enormously larger than the next largest city—in which a large percentage of the urban population is concentrated and in which most of the nation's industries, nonagricultural job opportunities, higher-order urban services and facilities, and modern infrastructure and establishments are found.

Although not all developed economies have rank-size distributions of cities—those of small physical size tend to have higher levels of primacy—and not all underdeveloped countries have primate city settlement systems, there does seem to be some relationship between economic development and city-size distribution. In his review of city-size distributions in developing countries, Chetwynd notes that high primacy "tends to be associated with conditions of underdevelopment, e.g., export orientation, low per-capita GNP, high proportion of labor force in agriculture, and rapid population growth."[20]

Some theorists argue that the dominance of a primate city or the absence of an articulated and integrated system of central places obstructs the emergence of a sectorally and geographically balanced pattern of development. Johnson argues that the "varied hierarchy of central places has not only made possible an almost complete commercialization of agriculture but facilitated a wider spatial diffusion of light manufacturing, processing and service industries ... [and provided] employment of a differentiated variety," in both Western industrial countries and in more advanced economies in the developing world.[21] Poor countries without such spatial systems cannot achieve widespread development and reduce regional and urban-rural disparities because, lacking an integrated system of intermediate cities, small towns, and market centers, farmers cannot sell their surpluses, obtain inputs, modernize their technology, and adapt products to consumer demand. Nor can they easily obtain the services needed to make living in rural areas desirable.

In reaction to the overconcentration of investment in the largest metropolitan areas, some theorists have argued for shifts in investment allocation away from cities to rural areas and small towns. Friedmann supports an agropolitan development strategy: Programs would be integrated within rural districts of about 50,000 people and linkages with larger cities would be discouraged or developed only selectively.[22] But others contend that such a strategy overlooks the crucial economies of scale offered by cities of larger size and their potentially beneficial stimuli. Even those who formerly argued that the largest metropolitan centers were optimal for stimulating rapid economic growth now admit, as does Mera, that "either approach is insufficient." He argues that "the conventional city size approach neglects the spatially differentiated impact of a city on its surrounding areas, and the agropolitan approach neglects the benefit of agglomeration economies within a city."[23]

In reality, no particular spatial pattern is, a priori, universally optimal or desirable. The value of one pattern of settlement over another can only be assessed in relation to national policy objectives and development goals. Richardson is correct in pointing out that "if interregional equity is an important objective, a dispersed urban system with large regional metropolises in each region might be regarded as highly efficient." But if rapid industrialization is the goal, the size and spatial dispersion of cities may not be important; the objective is to achieve "industrial targets in urban areas with locational advantages, regardless of their size and location." In countries where agricultural and rural development are at least of equal priority to industrialization, the creation of a central place hierarchy, with a "punctiform network of rural service centers up to the regional metropolis, will determine the efficiency of the settlement pattern."[24] Thus for those countries pursuing a policy of equitable economic growth in which both urban and rural populations can benefit from development, concentration of investment in the largest metropolitan areas is likely to be undesirable and widespread distribution in rural areas is likely to be ineffective. A concept of more balanced spatial development—deconcentrated urbanization—has emerged; it underlines the importance of moving away from the highly skewed distribution of urban population and resources found in primate city systems and toward a more diffuse pattern of urbanization. Friedmann has argued that although "balanced development" also remains an elusive phrase, it implies a lessening or prevention of locational disparities in income and wealth. "No rigid, mathematical balance is intended," Friedmann asserts. "What is meant instead is a sense of systematic inter-relation between countryside and city in which their notorious differences in levels of living and opportunity will become progressively less pronounced."[25]

In some countries or regions, this implies the need for a deconcentrated, hierarchical pattern of urbanization; in others, even more diffuse patterns may be appropriate. Hackenberg argues that where secondary cities have stimulated the economies of lower-order urban centers, such as in the Southern Mindanao resource frontier of the Philippines, there has been a reduction of intraregional urban-rural income disparities and of interregional inequities in living standards nationally. Hackenberg contends that the strengthening of secondary cities and the linkages between them and smaller cities and towns can stimulate equitable, bottom-up development without necessarily creating a classical urban hierarchy or lognormal rank-size distribution.[26]

Ultimately, however, these theories of spatial development have had much less impact on emerging policies for secondary city development than have the pragmatic problems that governments in developing countries face in generating economic growth with social equity. The increasingly severe social, economic, physical, and political problems associated with rapid growth of the largest metropolitan areas, together with the worsening poverty in rural areas, impel planners and policymakers to search for new approaches to spatial development.

Reversing Polarization

Development planners began to realize at the end of the 1970s that no policy aimed at alleviating poverty or promoting equitable economic growth could ignore the pace of urbanization in the developing world or the continued concentration of urban population in the largest cities. From the 1950s, urbanization in developing countries has been at least equal to if not more rapid than in Western industrial societies during their peak periods of urban growth. Moreover, developing countries have reached their present levels of urbanization much more rapidly than did industrialized nations. Since 1950 the 4 percent average annual growth of urban population in developing countries has been more than double that of Western countries. Although the percentage of population living in urban areas in less developed nations is well below that of North America and Western Europe, the absolute number is slightly higher, and is projected by the United Nations to be more than double that of the now-industrialized countries by the end of the twentieth century. By that time, the developed countries are expected to have about 1 billion people living in urban areas, and the developing countries about 2.1 billion.[27]

Although the pace and scale of urbanization differ among regions, rapid growth in urban population is characteristic of large parts of the developing world. Unlike in Western countries, urban and rural populations in developing nations are growing rapidly at the same time. In East and Southeast Asia (excluding Japan) there were more people—596 million—living in cities in 1980 than in North America and Europe combined. About 133 million Africans and 240 million Latin Americans were also living in urban places. By the end of the 1990s the urban populations of Southeast Asia and of Africa are expected to triple, and those of South Asia and Latin America to double. Given current trends, more than 60 percent of the population of Northern Africa and the Middle East will be living in urban areas by the end of the 1990s, as will be three-quarters of the populations of South and Central America and East Asia.

Migration from rural to urban areas continues to be high even as governments seek to increase rural productivity and income. About half of the urban population increase in developing countries is attributed to rural migration. Countries with severe rural poverty such as Indonesia, Pakistan, Bangladesh, and India—are expected to experience strong rural migration in coming decades. The World Bank predicts that in these countries "intensive pressure on land—the rural population per square kilometer of arable land is expected to double in the next 25 years—which, even if it leads to a small percentage of outmigration from rural areas, will lead to massive growth in the cities."[28] Indeed, rural development may accelerate emigration rather than slow it. Rural migrants, for the most part, join the growing numbers of urban poor.

Thus many governments in developing countries have become dissatisfied with the spatial distribution of their population. Of 116 governments surveyed by the United Nations in 1978, 68 reported that they were strongly dissatisfied with current population distribution and an additional 42 expressed partial dissatisfaction. Only 6 governments considered current patterns acceptable. In Africa, the complaint was over extreme dispersion and lack of sufficient agglomeration to provide services efficiently. In Latin America and Asia dissatisfaction arose from the increasing concentration of people in a few large metropolitan centers. In Central America and the Middle East the concern was with finding ways to achieve more balanced distribution without promoting the spatial polarization found in many industrializing nations.[29]

RAPID GROWTH OF THE LARGEST CITIES

One of the strongest reasons for increased interest in secondary city development is the unrelenting growth of the largest metropolitan centers. Developing nations will soon surpass industrial countries in the number and size of large metropolitan areas. In 1950, only 3 cities in developing countries had populations of 4 million or more. By 1975 there were 17 such cities, and United Nations demographic studies predict that by the end of the 1990s there will be 61 cities in the developing world with more than 4 million people, compared to about 25 in the West. If present growth trends hold, by the end of the century, 21 of the world's 30 largest metropolitan centers will be in developing nations. The portion of urban population living in these massive urban centers has tripled in 25 years, from a little more than 5 percent to about 15 percent. The combined population of the largest metropolitan areas in developing nations will exceed 323 million by the end of the 1990s. Mexico City is expected to become the world's largest metropolitan agglomeration with 31 million people, followed by São Paulo, Brazil, with 26 million. Shanghai and Peking are expected to reach 20 million people each, Rio de Janeiro to surpass 19 million, and Bombay, Calcutta, and Jakarta to grow to more than 16 million. Seoul, Cairo, Madras, Manila, and Buenos Aires, along with Bangkok, Karachi, Delhi, and Bogota, will exceed 12 million, and Teheran, Baghdad, and Dacca are expected to reach or surpass 10 million in population.[30]

This scale of agglomeration far exceeds that of metropolitan areas in Western countries during their post-World War II period of industrial growth. Developing nations, however, have far fewer resources to cope with problems of massive urbanization. The largest metropolitan centers in developing nations at the end of this century will be 2 or 3 times the size of the largest metropolises in industrialized nations in 1950. Moreover, metropolitan centers of 1 million or more population have been growing rapidly in nearly every developing region (see Table 1.1). There were 31 cities of a million or more people in developing nations in 1950, containing about 22 percent of the urban population. The number tripled to 90 in 1975, and at that time they accounted for one-third of the urban dwellers. By 1990, Africa's "million cities" are expected to increase from the 2 that existed in 1950 to 37, claiming about 38 percent of the urban population. Latin American cities of a million or more people will have increased from 7 in 1950 to 42 in 1990, and will contain 48 percent of the urban residents in the South and Central Americas. East Asia will have 61 such cities with 40 percent of the urban population and South Asia will have 65 cities of a million or more residents with 38 percent of the urban population. If present demographic trends hold, there will be about 284 metropolitan centers of a million or more people in developing nations by the end of the 1990s, with a combined population of nearly 1 billion.[31]

TABLE 1.1 Population and Projected Size of Cities of One Million or More Population in Developing Countries: 1950-2000

	1950	1960	1970	1980	1990	2000
Less developed countries[a]						
population[b]	62.345	113.446	192.932	339.373	595.130	931.835
number of cities	31	50	74	118	198	284
Regions						
Africa						
population[b]	3.503	7.482	15.415	36.485	83.363	154.158
number of cities	2	44	8	19	37	57
Latin America						
· population[b]	17.276	30.988	56.383	101.301	164.167	232.158
number of cities	7	11	17	27	42	57

East Asia						
population[b]	33.771	62.117	90.495	131.910	191.270	261.554
number of cities	14	24	31	42	61	82
South Asia						
population[b]	19.360	33.267	58.837	105.879	199.085	328.163
number of cities	11	16	23	36	65	95
Oceania						
population[b]	3.136	4.021	5.021	7.185	11.903	13.427
number of cities	2	2	2	3	6	6

SOURCE: Compiled from United Nations, *Patterns of Urban and Rural Population Growth*, Population Studies 68 (New York: United Nations Department of International Economic and Social Affairs, 1980) Tables 20, 22.

a. Does not equal total of regions.
b. In millions.

HIGH LEVELS OF POPULATION
AND RESOURCE CONCENTRATION
IN PRIMATE CITIES

It is not merely the pace of urbanization that presents growing problems, but also the pattern of urban population concentration and the social and economic inequities usually associated with spatial polarization with which developing countries will have to cope in the future. In many developing countries, the middle levels of the urban hierarchy—including secondary cities—that could absorb more migrants and create a more balanced distribution of urban population are extremely weak. In those with primate city spatial structures the largest metropolis has grown to such a size and level of wealth that it dominates the nation's settlement system and economy.

Countries with polarized urban systems include those that have already become highly urbanized and those that are beginning to industrialize, such as South Korea, as well as those highly urban countries, such as Egypt, Lebanon, El Salvador, Panama, Peru, and Chile, in which economic growth has been sluggish and the overwhelming majority of the people remain poor (see Tables 2.2-2.5, in Chapter 2). [32] In Korea, nearly half of the population was urbanized in 1980 and more than 40 percent of the urban population was concentrated in Seoul. United Nations estimates place more than 64 percent of Panama's urban population in its national capital, but the government's census counts for 1978 claim that 80 percent are living in Panama City and that by the end of the century two-thirds of the country's entire population may be living in a single metropolitan area. [33] In Egypt, about 38 percent of the urban dwellers are in Cairo and another 15 to 20 percent are concentrated in Alexandria. Lebanon's population is more than 40 percent urbanized, with 60 percent of the urban residents in Beirut.

Primate city structures are also characteristic of a number of countries that remain predominantly rural with a relatively small, but highly concentrated, urban population. Although less than one-third of their populations live in urban areas, these countries have seen 1 or 2 metropolitan centers grow to enormous size. Well over 45 percent of the Philippines' urban population is living in metropolitan Manila and more than 68 percent of Thailand's is concentrated in the Bangkok region. In Latin America, more than 43 percent of Bolivia's urban population is concentrated in La Paz; 64 percent of Costa Rica's is in San Jose; and 65 percent of Jamaica's is in Kingston. More than 52 percent of the Uruguayan urban population was in that country's capital city in 1980. Similar patterns appeared even in the poorest and most rural countries of Africa. From half to nearly 90 percent of the urban populations of Mozambique, Burundi, Senegal, Guinea, Angola, Uganda, Kenya, and Tanzania live in single cities within those countries.[34]

Primate cities and large metropolitan centers also have concentrations of national resources and social overhead capital vastly greater than their share of national population, and from which only a small percentage of the nation's people obtain direct benefits. The metropolitan centers grew so quickly and to such a large scale in part because they have been favored locations for investments, especially those related to export production, capital-intensive manufacturing, and modern infrastructure. In Indonesia, for instance, almost 60 percent of foreign investment and 26 percent of private investment during the 1970s was made in Jakarta, which contained only about 4 percent of the nation's population. Central government investments in Jakarta were more than double the amount made in the province receiving the next highest allocation.[35]   Although central Manila has only about one-fourth of the population of the Philippines, the bay area of central Manila alone accounts for more than 72 percent of the nation's manufacturing firms, 80 percent of all manufacturing employment and production, and 61 percent of the nation's hospital beds. It produces and consumes more than 80 percent of the nation's electrical power and generates 65 percent of the country's family income.[36] Similarly, Bangkok has received about 65 percent of the annual investment in construction in Thailand and accounts for nearly three-fourths of all commercial bank deposits in the country.[37] South Korea's two major metropolitan centers clearly dominate the rest of the nation in social, economic, and infrastructure investment. In the mid-1970s over 60 percent of the nation's medical services were concentrated in Seoul and Pusan, with Seoul having the largest portion. More than 60 percent of educational services, 71 percent of wholesale establishments, and more than half of transportation services were located in these two cities.[38] Similar patterns of concentration of productive activities, industrial employment, transport services, and modern infrastructure and facilities appear in metropolitan centers in many Latin American and Near Eastern countries.

Although many of these countries have enacted laws aimed at decentralizing economic activities, social services, and commercial and public facilities from primate cities, the policies have not been successful. In part, their failure is due to the paucity of secondary cities of sufficient size and diversity to support high population threshold activities outside of the national capital. Many countries in the developing world have relatively few secondary cities, and their rural populations are largely scattered in small villages. The secondary cities that do exist, as will be seen later, have neither played a strong role in absorbing rural-to-urban migration nor developed the financial, administrative, and economic capacities to act as countermagnets to the primate city. In some countries secondary cities are not widely distributed in a pattern that can generate more widespread and balanced urban growth. Thus a number of governments, such as Thailand's, are seeking to combine policies for controlling the growth of the primate city with those for stimulating the development of secondary urban centers that can become more viable locations for activities that have nowhere else to locate but in the primate city. The Thai strategy, outlined in the National Economic and Social Development Plan for the 1980s, involves two major objectives: first, to limit the growth of the metropolitan area "at some appropriate level so as to check urban congestion and improve the existing environment," and second, to develop regional growth centers and upgrade small and medium-sized towns to populations of from 100,000 to 300,000, a size that would enable them to provide a wide variety of manufacturing, agricultural, commercial, and social services.[39]

Alleviating Problems in Larger Cities

The growth of massive metropolitan areas and primate cities has created serious economic and social problems with which most developing countries do not have the resources to cope directly. The largest cities in Latin America are experiencing serious housing, transportation, pollution, employment, and service supply problems. High levels of underemployment among squatters and recent migrants maintain these people in poverty. Difficulties in extending and maintaining existing sewer, water, and drainage systems and utility services often create health and sanitation problems in densely populated squatter areas, and the strains on existing social, health, and educational services become more severe with population growth.[40] Similar problems plague the largest cities in the Middle East and Asia.[41]

The costs of meeting social needs in the largest metropolitan centers are, of course, impossible to calculate with great accuracy, but almost all attempts have produced extremely high estimates that are beyond the present or potential capacity of developing nations to pay. Using rough but conservative estimates, Unikel found in 1970 that if

Mexico City reached a projected population of 21 million by 1990, the national government would have to spend the equivalent (in 1970 dollars) of about $3.2 billion a year over the next 20 years to settle new migrants within the city and to supply the existing population with minimal housing, water, electricity, transportation, and health and educational services. Unikel concluded that this amount—which does not include the cost of reducing deficits in housing, services, and jobs over the period—could hardly be obtained. It is equivalent to the cost of establishing a city comparable in size to Guadalajara, Mexico's second largest metropolis, every year.[42] Prakash estimated the costs of providing even minimum facilities in Asia's largest metropolitan centers at between $5.2 and $7.5 billion a year during the 1980s and up to $11 billion a year during the 1990s. He notes that services are not only strained in Asian cities, but that they are deteriorating rapidly.[43] By building up the capacity and increasing the number of secondary cities, many governments hope to relieve population pressures on the largest metropolises and contain the growth of secondary cities to sizes that are more manageable.

Reducing Regional Inequities

Despite all of these problems, living conditions for most people are often perceived to be better in the largest urban centers than in smaller towns or rural areas. Per capita income of people living in Bangkok, for example, is 4 times higher than the average in Thailand's rural areas, and Bangkok's average is 292 percent higher than that of the country as a whole.[44] Average per capita disposable income in San Salvador is twice as high as in El Salvador's rural areas and 88 percent higher than in the country's other urban centers.[45]

Even when adjustments are made for differences in costs of living, residents of major metropolitan areas, on average, are far better off than people living in the rest of the country. For example, measurements of the physical quality of life, based on health, educational, and social indicators, show that people living in Tanzania's capital city have far better conditions than rural people.[46] Similar conclusions were drawn from analyses in the Philippines: 96 percent of the households in Manila have electricity, compared to 28 percent in the country as a whole; 83 percent in Manila have sanitary water supplies, compared to about 40 percent in the rest of the country; more than half in Manila have flush toilet facilities, compared to 17 percent outside of Manila. Life expectancy is higher in metropolitan Manila than anywhere else in the country, infant mortality is lower, and a substantially smaller percentage of the metropolitan area's population is living in abject poverty than in any other place in the Philippines.[47]   Thus some planners argue that developing secondary cities is one means of spreading the benefits of urbanization to larger numbers of people and of reducing interregional disparities.[48]

Stimulating Rural Economies

If the serious problems of large-scale urban concentration have been one motivation to promote secondary city development, the contentions that urban and rural development are inextricably related, that intermediate cities can play important roles as catalysts for rural development, and that a system of secondary cities can be important in achieving economic growth with social equity have been far more influential. World Bank analysts contend that secondary city development can promote more equitable economic growth in rural regions aside from whatever impact it has on slowing growth in large metropolitan areas. In a study for the World Bank, Richardson argues that these benefits can include commercialization of agriculture, provision of better services to residents in rural regions, national spatial integration, diffusion of social and technical innovations from the major metropolitan areas and from abroad, the decentralization of job opportunities, and, "most important of all, the more equitable distribution of welfare (among urban areas and among regions) resulting from an intermediate city strategy."[49] World Bank analysts emphasize the importance of planning to shape the urban system before a country has reached high levels of urbanization, noting that "excessive urban concentration is difficult to correct once it has occurred."[50]

Most rural countries lack sufficient numbers of intermediate cities to stimulate agricultural economies. These include India, Pakistan, and Indonesia, which are basically rural countries but which also have a few large metropolitan centers, and most countries in Africa and the Near East, which have no significant urbanization. In Tanzania, Mauritania, Nigeria, Upper Volta, Mali, and Yemen, the few urban centers that do exist are of such small size that population thresholds are too low to support the variety of services and facilities needed to stimulate commercial agriculture. In Mauritania, for example, the population of the largest city is less than 200,000 and the chief towns in rural regions range from a little more than 2,000 to less than 23,000 people.[51] Even in India, which has some of the largest cities in the world, the overwhelming majority of the population is scattered in small villages. Over 80 percent of India's nearly 550 million people live in villages and towns of less than 5,000, and nearly 93 percent of the rural settlements have less than 2,000 people.[52] The vast majority of the population in countries or regions without an integrated system of central places lack access to the services, facilities, and resources needed for rural development.

International assistance agencies are helping governments in Africa to locate infrastructure and marketing facilities in such a way as to strengthen the developmental capacities of secondary cities and market towns. In the Sudan, as in Yemen, Tanzania, and Upper Volta, the highly dispersed rural population remains largely unserved because rural settlements are too small to support most services and facilities needed to stimulate rural development. The cost of providing services to widely scattered populations is extremely high, and the use of such facilities is generally low. But studies of the Sudan point out that "without access to markets, inputs, education, health and other services, development of the traditional market sector cannot occur."[53] The Tanzanian government, recognizing that dispersal of population in very small settlements and on individual farm plots and weak linkages between rural areas and urban centers are serious obstacles to integrating large numbers of the rural poor into the national economy, has attempted to create a system of central places. Without a central place system, "the inability of rural farmers to market their goods wipes out any incentives offered by the government to increase production," one study points out. "The same problem exists with inputs supply. As a result many of the fertile areas are vastly underutilized."[54]

The value of secondary cities to rural development is underscored by the Thai government's plan for developing the Northeast Region. Studies of the region note that these cities "are in symbiotic relationship with the rural areas surrounding them. They are an important intermediate or ultimate market for agricultural production." Moreover, these studies found that secondary cities are also "the local source of agriculturally related capital goods and services upon which modernization of the subsistence system is dependent."[55]

In Panama, the government is encouraging a wider spatial distribution of population and assets, because it believes that "secondary growth centers must be developed to meet the structural shift to an off-farm, employment oriented economy." Studies of development needs in Panama point out that "besides increasing employment opportunities these centers must become complementary engines of development on a regional basis by providing nearby farm areas with necessary goods and services."[56] The Costa Rican government has discussed an extensive program of support for intermediate cities for many of the same reasons. Its assistance plan for 1982-1986 points out that "small and medium size urban places can fulfill important functions for their rural hinterlands and investment in towns and cities off the *Meseta Central* can therefore have a favorable impact on the rural sector as well. The inputs and support services which agricultural development requires can best be located in urban centers." A major problem in Costa Rica, as in many other developing nations, is that "the country is not sufficiently urbanized or spatially structured to support a dynamic and modern agricultural sector."[57]

Development strategies in these countries seek to build the capacity of secondary urban centers to provide services, facilities, and markets for agricultural products and to begin absorbing surplus labor as agriculture becomes more efficient. "There is almost universal agreement here that programs must be undertaken to generate productive non-agricultural employment," AID analysts in Indonesia point out. "Failure to achieve readily perceived success in this area could threaten the success of all other elements of the country's development effort."[58]

Other governments see the development of secondary cities as a way of creating more efficient agroprocessing and agricultural support industries in rural regions, providing greater access for rural people to commercial and personal services, increasing food production, and providing off-farm employment opportunities. These are among the most crucial problems facing the government of Indonesia. Migration of rural unemployed to the largest cities and the difficulties of increasing food production to keep pace with growing rural and urban populations both suggest, as Atmodirono and Osborn have pointed out, "important roles for middle cities in Indonesia's long range development—functions beyond the industrial and service ones ordinarily associated with urbanization." These authors argue that it is the intermediate cities in Indonesia that will have to bear the responsibility for absorbing large numbers of rural people, providing them with employment and services, and stimulating the expansion of food production on a large scale through application of urban-based technology and managerial methods.[59]

Along similar lines, it has been noted that in El Salvador the strengthening of secondary cities must be a key element of development policy because "recent demographic studies show that rural to urban migration, which until now has been relatively minor, will assume major proportions in coming years." Analysts conclude that "no matter how beneficial agrarian transformation may prove to be, industrialization (and inevitably urbanization) offers the only long range solutions to El Salvador's economic woes."[60]

Increasing Administrative Capacity

Many developing nations are giving higher priority to the middle level of the urban hierarchy for two other reasons. One is their desire to expand the capacity of secondary cities to perform service and production functions more efficiently and effectively. Many have low levels of administrative capacity, poor planning and management capability, inefficient service delivery programs, and low levels of revenue raising capacity, and are dependent on the central government for authority and financial resources to perform even basic functions. Thus they cannot easily fulfill their potential roles in absorbing rural migrants and stimulating the rural economy.

Secondary cities have become more important in Kenya's development strategy because it is predicted that the country's urban population will grow from 2.2 million in 1979 to 8.5 million at the end of the century and that "the socioeconomic problems attendant with this massive population transition will be most acute in Kenya's secondary cities."[61] Their governments must be able to take the initiative in generating employment, improving housing, expanding markets, and strengthening revenue-raising and taxing procedures. In Panama, Costa Rica, Honduras, and other Central American nations, new interest has been expressed in strategies to build up regional planning and administrative capacity, increase investment in infrastructure and housing in regional urban centers, and promote new sources of employment in larger towns and cities outside of the capital.

A closely related reason for developing more secondary cities is the growing recognition of their importance for decentralizing development planning and management. Experiments with administrative decentralization in the Sudan, Tanzania, and Kenya during the 1970s faltered for lack of cities of sufficient size, other than the national capital, to take up these responsibilities.[62] A system of geographically dispersed secondary cities seems essential for the decentralization of private investment as well. Long-term, consistent decentralization policies usually must precede deconcentration of private investment. The World Bank has noted that "to be effective, decentralization policies must be applied consistently over an extended period. A stop-go approach provides private investors with ambiguous signals and reduces their willingness to move from the largest cities."[63]

Reducing Urban Poverty
and Increasing Productivity

Finally, it has become increasingly clear in recent years that if equitable development consists, at least in part, of increasing the productivity and income of the poor, more attention must be given to their growing numbers in secondary cities. The inextricable relationship between urban and rural poverty is seen most clearly in intermediate cities, where urban and rural economies converge and where the problems of poverty and marginality "are more visible and often more acute."[64] World Bank estimates have placed 37 percent of the urban population in Africa below the poverty line ($100 per capita income). Nearly 26 percent survive on incomes too low for them to acquire even minimum basic human needs. In South Asia about 58 percent of the urban population lives in poverty and about 40 percent at or near subsistence levels. About one-fifth of those living in urban places in East Asia and the Pacific survive on subsistence incomes.[65] The problem of urban poverty cannot be ignored in policies aimed at generating economic growth with social equity, and cannot be dealt with effectively through rural development alone.

Recent surveys indicate that much of the urban poverty in developing countries is found in intermediate and smaller cities. In Costa Rica, for instance, "urban poverty, while concentrated in absolute numbers in San Jose, is actually worse in other areas of the country where almost 60% of all poor Costa Rican urban dwellers live."[66] High concentrations of urban poverty are found in the "ring cities" around the capital as well as in the secondary cities of Limon and Puntarenas and smaller towns such as Liberia. Studies of Ecuador note that the urban poor are "generally unprepared for city life and face great difficulties in becoming part of the productive market." Programs are needed that will help integrate the poor into the economies of secondary cities, where migrants are usually "self employed as peddlers, offer their services as unskilled construction workers or become personal services workers (e.g., domestics, dishwashers) earning just enough for the most pressing food needs." In coastal cities they join the squatters of the barrios and in the Sierra they crowd into rented rooms with no sanitary services; they live "in abject poverty, with a sense of helplessness and in an atmosphere of crime and vice."[67]

The Egyptian government is strengthening secondary cities to provide employment and meet the basic needs of the poor. "Short term priorities will be on programs . . . in the cities where nearly half of Egypt's population lives—and in particular attention must be given to targeting development programs . . . for urban middle classed and poor groups" in the secondary and smaller cities, AID analysts point out.[68]

In Kenya, the urban poor are expected to become more numerous as the urban population approaches a quarter of the country's total population by the end of this century.[69] Surveys of economic conditions in Pakistan point out that "while the prime importance of attacking rural poverty cannot be denied, the incidence of urban poverty appears somewhat disturbing and deserves attention." The capacity of secondary cities to employ the poor must be improved both because migrants continue to flow into urban places to escape rural poverty and because their concentration in the largest cities creates potentially severe social and political problems. "Despite higher incomes, the number of urban poor is higher than a similar figure for rural areas," observers contend. "Cities continue to amass slums at an ever increasing rate; and what is more important, acute awareness of poverty in urban areas has bred considerable discontent, expressed frequently in the form of street violence."[70]

Some analysts suggest that a secondary city development strategy, by emphasizing the importance of *place* rather than of *program,* is a "significant and innovative shift from the conventional emphasis on the need for planners to meet basic human needs largely by throwing money at them."[71] Hackenberg contends that a development policy aimed at creating a productive system of settlements can meet social needs more effectively by expanding employment opportunities and generating income for the poor, allowing them to demand and obtain the services they need. A productivity-directed policy, Hackenberg argues, must be concerned with "the economic and residential organization of the urban community," and thus it must be "place-oriented."[72] By increasing the productivity and income of the community, resources can be mobilized to provide and maintain services that people want.

Conclusions

Although the interest in developing secondary cities has emerged from a variety of theoretical and practical arguments about their potential roles in regional and national development, little is really known about secondary cities in developing countries. In the following chapters, their social, economic, and demographic characteristics, the factors that affect their growth and development, their functional strengths and weaknesses, and their potential for stimulating the economies of their hinterlands are explored.

NOTES

1. See, for instance, Michael Lipton, *Why Poor People Stay Poor: Urban Bias in World Development* (Cambridge, MA: Harvard University Press 1977); Michael P. Todaro and Jerry Stilkind, *City Bias and Rural Neglect: The Dilemma of Urban Development* (New York: Population Council, 1981).

2. Some of the evidence is reviewed by Alan B. Simmons, "Slowing Metropolitan Growth in Asia: Policies, Programs and Results," *Population and Development Review* 5 (March 1979): 87-104; S. U. Kim and P. J. Donaldson, "Dealing with Seoul's Population Growth: Government Plans and Their Implementation," *Asian Survey* 19 (July 1979): 660-673; W. A. Cornelius and R. V. Kemper, eds., *Metropolitan Latin America: The Challenge and the Response* (Beverly Hills, CA: Sage, 1978).

3. The relationship between investment allocation and regional disparities is discussed in detail in Dennis A. Rondinelli, "Regional Disparities and Investment Allocation Policies in the Philippines: Spatial Dimensions of Poverty in a Developing Country," *Canadian Journal of Development Studies* 1 (Fall 1980): 262-287.

4. See Dennis A. Rondinelli and Kenneth Ruddle, *Urbanization and Rural Development: A Spatial Policy for Equitable Growth* (New York: Praeger, 1978), the published version of a policy review study conducted for USAID as *Urban Functions in Rural Development: An Analysis of Integrated Spatial Development Policy* (Washington, DC: U.S. Agency for International Development, 1976).

5. U.S. Agency for International Development, "Urbanization and the Urban Poor," Policy Determination 67, Handbook 2, Memo 2:5 (Washington, DC: Author, 1976), p. 1.

6. World Bank, *Urbanization Sector Working Paper* (Washington, DC: Author, 1972), p. 4.

7. World Bank, *World Development Report, 1979* (Washington, DC: Author, 1979), p. 75.

8. Harold Lubell, *Urban Development Policies and Programs,* Working Paper for Discussion, Bureau for Program and Policy Coordination, Economic Development Division (Washington, DC: U.S. Agency for International Development, 1979), p. 3.

9. Ibid., p. 34.

10. Rondinelli and Ruddle, *Urbanization and Rural Development*; Dennis A. Rondinelli, "Balanced Urbanization, Spatial Integration and Economic Development in Asia," *Urbanism Past and Present* 9 (Winter 1979-1980): 13-29.

11. William C. Wheaton and Hisanobu Shishido, "Urban Concentration, Agglomeration Economies and the Level of Economic Development," *Economic Development and Cultural Change* (October 1981): 29.

12. Koichi Mera, "On the Urban Agglomeration and Economic Efficiency," *Economic Development and Cultural Change* 21 (January 1973): 309-324.

13. See Werner Z. Hirsch, "The Supply of Urban Public Services," in *Issues in Urban Economics,* ed. H. S. Perloff and L. Wingo, Jr. (Baltimore: Johns Hopkins Press, 1968), pp. 477-526; Colin Clark, *Population Growth and Land Use* (London: Macmillan, 1967).

14. Stanford Research Institute, "Costs of Urban Infrastructure for Industry as Related to City Size: India Case Study," *Ekistics* 20 (November 1969): 316-320.

15. Niles Hansen, "The Role of Small and Intermediate Size Cities in National Development Processes and Strategies" (Paper delivered at Expert Group Meeting on the Role of Small and Intermediate Cities in National Development, United Nations Centre for Regional Development, Nagoya, Japan, 1982), p. 1.

16. Ibid., p. 6.

17. Brian J. L. Berry, "Policy Implications of an Urban Location Model for the Kanpur Region," in *Regional Perspective of Industrial and Urban Growth: The Case of Kanpur,* ed. P. B. Desai et al. (Bombay: Macmillan, 1969), p. 207.

18. See Brian J. L. Berry and Frank E. Horton, *Geographic Perspectives on Urban Systems* (Englewood Cliffs, NJ: Prentice-Hall, 1970); see especially pp. 64ff.

19. Brian J. L. Berry, *Geography of Market Centers and Retail Distribution* (Englewood Cliffs, NJ: Prentice-Hall, 1967), p. 20.

20. Eric Chetwynd, Jr., "City Size Distribution, Spatial Integration and Economic Development in Developing Countries: An Analysis of Some Key Relationships" (Ph.D. diss., Duke University, 1976), pp. 84-85.

21. E.A.J. Johnson, *The Organization of Space in Developing Countries* (Cambridge, MA: Harvard University Press, 1970), pp. 28, 99, 171; see also Dennis A. Rondinelli and Kenneth Ruddle, "Integrating Spatial Development," *Ekistics* 43 (1977): 185-194; Dennis A. Rondinelli and Kenneth Ruddle, "Coping with Poverty in International Assistance Policy: An Evaluation of Spatially Integrated Investment Strategies," *World Development* 6 (April 1978): 479-498.

22. See John Friedmann, "A Spatial Framework for Rural Development: Problems of Organization and Implementation," *Economie Appliquee* 28 nos. 2-3 (1975): 519-544.

23. Koichi Mera, "City Size Distribution and Income Distribution in Space," *Regional Development Dialogue* 2 (Spring 1981): p. 105.

24. Harry W. Richardson, *City Size and National Spatial Strategies in Developing Countries,* World Bank Staff Working Paper 252 (Washington, DC: World Bank, 1977), p. 17.

25. John Friedmann, "The Active Community: Toward a Political-Territorial Framework for Rural Development in Asia," *Economic Development and Cultural Change* 29 (January 1981): p. 246.

26. Robert A. Hackenberg, "Diffuse Urbanization and the Resource Frontier: New Patterns of Philippine Urban and Regional Development" (Paper prepared for Expert Group Meeting on the Role of Small and Intermediate Cities in National Development, United Nations Center for Regional Development, Nagoya, Japan, 1982).

27. United Nations, Department of International Economic and Social Affairs, *Patterns of Urban and Rural Population Growth,* doc. no. ST/ESA/SER.A/68 (New York: United Nations, 1980), p. 11.

28. World Bank, *The Task Ahead for Cities in the Developing Countries,* World Bank Staff Working Paper 209 (Washington, DC: Author, 1975), p. 11.

29. See United Nations, Department of International Economic and Social Affairs, *World Population Trends and Policies: 1979 Monitoring Report,* vol. 2 (New York: United Nations, 1980), pp. 40-41.

30. Ibid., pp. 48-54.

31. Ibid., pp. 53-58.

32. Data for 1970 are drawn primarily from Kingsley Davis, *World Urbanization 1950-1970,* vol. 1 (Berkeley: University of California, Institute of International Studies, 1969).

33. U.S. Agency for International Development, *Country Development Strategy Statement [CDSS], Panama, FY 1982-1986* (Washington, DC: Author, 1980), pp. 4-5.

34. Data drawn from Davis, *World Urbanization,* and from Tables 2.2-2.5 of this study. See also the summary of primacy indexes for developing countries calculated by Bertrand Renaud, *National Urbanization Policies in Developing Countries,* World Bank Staff Working Paper 347 (Washington, DC: World Bank, 1979), p. 31.

35. Abukasan Atmodirono and James Osborn, *Services and Development in Five Indonesian Middle Cities* (Bandung: Bandung Institute of Technology, Center for Regional and Urban Studies, 1974), pp. 14-15.

36. Republic of the Philippines, National Economic and Development Authority, *Manila Bay Region in the Philippines,* Interim Report 3 (Manila: NEDA, 1973).

37. Jeff Romm, *Urbanization in Thailand* (New York: Ford Foundation, 1975), p. 7.

38. Republic of Korea, [Long Range Planning for Urban Growth to the Year 2000: Data Collections], vols. 1-2 (Seoul: Ministry of Construction, 1980).

39. Government of Thailand, *The Fourth National Economic and Social Development Plan, 1977-1981* (Bangkok: National Economic and Social Development Board, 1977), pp. 224-228.

40. A summary of metropolitan problems in Latin America can be found in Wayne A. Cornelius, "Introduction," in Cornelius and Kempter, *Metropolitan Latin America,* pp. 7-24.

41. See Vincent J. Costello, *Urbanization in the Middle East* (London: Cambridge University Press, 1977); M.H.P. Roberts, *An Urban Profile of the Middle East* (New York: St. Martin's, 1979).

42. Luis Unikel, "Urbanization in Mexico: Process, Implications, Policies and Prospects," in *Patterns of Urbanization: Comparative Country Studies,* vol. 2, ed. S. Goldstein and D. F. Sly (Liege, Belgium: International Union for the Scientific Study of Population, 1977), pp. 559-560.

43. Ved Prakash, *A Study of Fiscal Policy and Resource Mobilization for Urban Development* (Manila: Asian Development Bank, 1977).

44. U.S. Agency for International Development, *CDSS, Thailand, FY 1982* (Washington, DC: Author, 1980), p. 74.

45. U.S. Agency for International Development, *CDSS, El Salvador, FY 1982-1986* (Washington, DC: Author, 1980), p. 2.

46. U.S. Agency for International Development, *CDSS, FY 1981: Tanzania* (Washington, DC: Author, 1979), p. 2a.

47. See V. B. Paqueo, "Social Indicators for Health and Nutrition," in *Measuring Philippine Development,* ed. M. Mangahas (Manila: Development Academy of the Philippines, 1976), p. 81; Social Research Associates, *An Analytical Description of the Poor Majority,* Project Report I-B (Manila: NEDA, 1977); U.S. Agency for International Development, *CDSS, Philippines, FY 1981* (Washington, DC: Author, 1979).

48. U.S. Agency for International Development, *CDSS, Costa Rica, FY 1982-1986* (Washington, DC: Author, 1980), p. 42; idem, *Colombia: Urban-Regional Sector Loan,* Sector Loan Paper AID-DLC/P-966 (Washington, DC: Author, 1975); idem, *Urban-Regional Sector Analysis* (Bogota: USAID Mission, 1972).

49. Richardson, *City Size and National Spatial Strategies,* p. 11.

50. World Bank, *World Development Report, 1979,* p. 77.

51. Cited in U.S. Agency for International Development, *Draft Environmental Report on Mauritania* (Washington, DC: Library of Congress, Science and Technology Division, 1979), pp. 1-2.

52. See Om Prakash Mathur, *The Problem of Regional Disparities: Analysis of Indian Policies and Programmes* (Nagoya, Japan: United Nations Centre for Regional Development, 1975).

53. U.S. Agency for International Development, *CDSS FY 1981: Sudan* (Washington, DC: Author, 1979), p. 8.

54. U.S. Agency for International Development, *CDSS FY 1981: Tanzania* (Washington, DC: Author, 1979), p. 31.

55. U.S. Agency for International Development, *CDSS, Thailand*, p. 52.

56. U.S. Agency for International Development, *CDSS, Panama*, p. 49.

57. U.S. Agency for International Development, *CDSS, Costa Rica*, p. 42.

58. U.S. Agency for International Development, *CDSS, Indonesia, FY 1982* (Washington, DC: Author, 1980), p. 33.

59. Atmodirono and Osborn, *Services and Development*, p. 85.

60. U.S. Agency for International Development, *CDSS, El Salvador*.

61. U.S. Agency for International Development, *CDSS, Kenya, 1982-1986*, (Washington, DC: Author, 1980), p. 11.

62. See Dennis A. Rondinelli, "Government Decentralization in Comparative Perspective: Theory and Practice in Developing Countries," *International Review of Administrative Sciences* 47, no. 2 (1981): 133-145; idem, *Administrative Decentralization and Area Development Planning in East Africa: Implications for United States Aid Policy*, Occasional Paper 1 (Madison: University of Wisconsin Regional Planning and Area Development Project, 1980).

63. World Bank, *World Development Report, 1979*, p. 78.

64. Lubell, *Urban Development Policies*, p. 3.

65. Ibid, p. 23.

66. U.S. Agency for International Development, *CDSS, Costa Rica*, p. 41.

67. U.S. Agency for International Development, *CDSS, Ecuador, FY 1981*, (Washington, DC: Author, 1979), pp. 4-5.

68. U.S. Agency for International Development, *CDSS, Egypt, FY 1982-1986* (Washington, DC: Author, 1980), p. 3.

69. U.S. Agency for International Development, *CDSS, Kenya*, p. 31.

70. U.S. Agency for International Development, *CDSS, Pakistan, FY 1981* (Washington, DC: Author, 1979), p. 7.

71. Robert A. Hackenberg, personal communication, June 24, 1982.

72. Ibid., p. 2.

CHAPTER 2

DEFINING "SECONDARY CITIES"

Inevitably, the range of cities that constitutes the secondary level in an urban hierarchy varies among countries, depending on their patterns of urban settlement, levels of development, and economic structures. But some general notion of what a secondary city is must precede attempts at comparative analysis. The most convenient and frequently used criterion is relative population size. The number of residents alone, however, cannot adequately define a secondary city. Population density, physical size, the proportion of the labor force engaged in nonagricultural occupations, the mix and diversity of functions located within a city, its physical characteristics, and its relationships with other cities and towns must all be used to refine demographic criteria.

Most analysts expect secondary cities to perform central place functions; that is, their economic and social activities—and thus the cities themselves—must serve people living outside their boundaries. Others argue that the functions must also be interactive—that secondary cities should connect, transfer, and disseminate, and that they should serve as channels for the flow of goods and services, mediate social relationships, and diffuse modernization influences. Atmodirono and Osborn contend, for instance, that secondary cities in Indonesia "possess a distinct middle function in development policy flows between the center (Jakarta and the central government) and the smaller regional settlements in economic, social, governmental and generally modernizing

An earlier version of material appearing in this chapter was published by the author in an article appearing in *The Third World Planning Review*, Vol. 4 (1982), and is reprinted with the permission of Liverpool University Press.

respects."[1] Osborn argues further, in his analysis of secondary cities in Malaysia and Indonesia, that the term also connotes "functional intermediacy in the flows of power, innovation, people and resources among places."[2]

Determining functional intermediacy is quite difficult, however, and requires detailed, longitudinal analysis of individual cities. Little aggregate information usually exists about functional characteristics of cities or flows of people and resources among them. Thus analysts usually fall back on data that are readily available, and use population size as an initial criterion.

Population data can be both convenient and initially useful for identifying secondary cities because a good deal of empirical evidence from both developed and developing countries shows a positive correlation between city size and functional complexity.[3] If, in fact, there is a strong relationship between population size and the variety of functions found in cities, then population at least provides a starting point for identifying the universe of cities to be examined in more detail.

This study will focus on the range of cities with populations of at least 100,000, up to but not including the largest city in the country. This classification is similar to Lubell's, which divides urban places into three levels:

a) the *metropolis*—a large city, usually the national capital of a small country (e.g., Lima in Peru) or a major regional capital in a large country (e.g., São Paulo and Rio de Janeiro in Brazil; Calcutta, Bombay and Madras in India); b) *secondary centers*— small cities ranging in population from 100,000 to 2.5 million or more; and c) the rural-urban interface—*small cities* or rural growth centers ranging, according to country context, downward in population from 100,000.[4]

Although this definition of secondary cities—urban places other than the largest city with a population of 100,000 or more—may exclude, in some developing countries, smaller towns that governments may consider to be secondary urban centers and that indeed have urban characteristics, it offers a high probability of encompassing most of the cities that perform essential urban functions. It is assumed that in most developing countries, towns of less than 100,000 population are predominantly agricultural and rural service centers. Some evidence to support this contention will be offered later.

The upper limit of the range of secondary cities must also vary. In highly urbanized countries the largest city may have 10 or more million people and secondary metropolises may reach 2 or 3 million in population. In countries with low levels of urbanization the largest city may not have yet reached 1 million. There is, however, a wide gap in population size between the largest city and secondary urban centers in many developing nations that makes them easy to identify in practice. This is especially true in countries with primate city settlement systems because the primate city, by definition, has a substantially larger portion of urban population than the next 2 or 3 largest cities. In these countries, the second-level metropolitan centers can easily be distinguished from the largest and they in turn are distinct in size from smaller cities. In Egypt, for example, the Cairo metropolitan area, with more than 6.5 million people, has almost 3 times the population of the next largest city, Alexandria, and is nearly 60 times the average size of all other cities with more than 100,000 people. In Mexico, there is a very distinct size difference between Mexico City—with over 15 million people—and the second largest metropolitan centers of Guadalajara and Monterrey, with 2 to 3 million people each. The average size of all other secondary cities in Mexico is less than 300,000.[5] Similarly, Bogotá was 4 times the size of Medellín and Cali in 1975, and more than 16 times the size of other intermediate cities in Colombia. Jakarta was at least 3 times larger than Bandung and Surabaya, and 9 times larger than the average secondary city in Indonesia. In South Korea, the Seoul metropolitan area had about 7 million people in 1978, a population 3 times larger than that of Pusan and 5 times larger than that of Taegu, the country's next largest cities.

Demographic Characteristics

However, any attempts to analyze the population of secondary cities in developing countries quickly meet severe data limitations. The primary sources of information, national censuses, are always somewhat out of date, and they usually provide little information other than population size for secondary cities. In some countries, data are not disaggregated for smaller cities and are reported only for provinces or districts. An even more troublesome problem is that some cities may be added to or subtracted from a size class simply because their physical boundaries have been changed between census surveys. Thus it is not unusual to find discrepancies in the number and size of cities in various size classes reported by international agencies and national governments. Despite these limitations, data compiled by the United Nations and other organizations provide some useful insights on general population growth and distribution trends for secondary cities and on their positions in the urban systems of Third World countries.

Demographic data show that although secondary cities have been growing in both number and population since 1950, their capacity to absorb urban population increases has been weak in most developing nations. In most regions of the developing world they have grown more slowly than both smaller urban places and the largest metropolitan center. Few cities have grown larger than 100,000 population in Africa, Central America, and the poorest countries of Asia and South

America. Secondary cities in most parts of the developing world have absorbed a much smaller share of rural migrants than the largest cities over the past quarter century. In some countries, the secondary cities' share of urban and total population has been declining. Yet, despite their relatively weak position in the urban hierarchy of most countries, the number of secondary cities increased to over 640—with a population of more than 327 million—in 1980. In sheer number and population size they are an important component of the urban settlement system in developing nations. Demographic characteristics and trends differ widely among countries, of course, and generalizations must be tempered by an examination of those differences.

Secondary cities in developing countries have been growing rapidly in both number and population since 1950. United Nations demographic surveys indicate that the number of secondary cities more than doubled between 1950 and 1980 (see Table 2.1). In Africa, the Middle East, and South America their number more than tripled. By 1980 there were 81 secondary cities in Africa, compared to 22 in 1950; the numbers increased in the Middle East from 19 to 66, in South America from 34 to 110, and in Central America from 13 to 37. In Asia the number of secondary cities rose nearly 65 percent, to about 350. Countries recording sharp increases in their secondary cities since 1950 include: Egypt, where the number grew from 5 to 14; Nigeria, from 5 to 20; South Korea, from 7 to 17; India, from 76 to 144; Pakistan, from 8 to 20; and the Philippines, from 6 to 15. In Turkey, Brazil, Mexico, and Colombia, the number of secondary cities increased threefold over the 30-year period (see Tables 2.2-2.5).

The population of secondary cities is estimated to have increased even more dramatically. In Africa it rose from about 18 million to about 30 million between 1970 and

TABLE 2.1 Number, Population, and Growth Rates of Secondary Cities in Less Developed Regions: 1950, 1970, and 1980

Region	Number			Secondary Cities Population (in millions)		% Increase in Population 1970-1980*
	1950	1970	1980*	1970	1980*	
Africa	22	74	81	18.203	30.024	64.9
Middle East	19	56	66	14.068	24.617	74.9
Central America	13	36	37	10.153	16.847	65.9
South America	34	105	110	32.263	51.988	61.1
East and South-east Asia	213	335	350	150.397	204.079	35.7
All Regions	301	606	644	225.084	327.555	45.5

SOURCE: Compiled from United Nations, *Patterns of Urban and Rural Population Growth*, Population Studies 68 (New York: United Nations Department of International Economic and Social Affairs, 1980), Table 48.

* Projected on the basis of the number and population of cities of 100,000 or more population in 1970.

1980, in the Middle East it grew from 14 million to 24 million, in Central and South America from 42 million to nearly 69 million, and in East and South Asia from 150 million to 204 million. The average increase in population living in secondary cities in all developing regions was over 45 percent between 1970 and 1980, ranging from 36 percent in Asia to 75 percent in the Middle East.

In African countries the populations of secondary cities grew at an annual average of nearly 7 percent during the 1960s and 6.5 percent during the 1970s. In the Middle East secondary city populations grew by an average of 9 percent a year during the 1960s; the rate of growth increased to slightly more than 10 percent a year during the following decade. Secondary city populations grew by more than 11 percent in Central America and 15 percent a year in South America during the 1960s, and by 6.5 and 6.1 percent, respectively, during the 1970s. The most rapid growth of secondary city populations in Asia occurred during the 1950s; but these cities were still growing by about 4.2 percent a year during the 1960s and by nearly 3.6 percent annually during the following decade. In nearly all countries, the population growth rate of secondary cities exceeded that of the nation as a whole.

Relatively, however, secondary cities have been growing more slowly than cities in larger and smaller size categories. Despite the rapid growth in the number and population of secondary cities over the past 30 years, they have been growing relatively slowly compared to the largest cities, and to smaller towns, in most developing nations. Only in those countries in Latin America and Asia where the largest cities have already reached enormous sizes have the populations of secondary cities grown faster. In Latin America, the populations of secondary cities grew about 8 percent faster than the total population of the largest cities. In Asia the increase in their population was nearly double that of the largest cities.

(text continues on page 59)

TABLE 2.2 Number, Population, and Growth Rates of Secondary and Largest Cities in Developing Countries of Africa: 1970 and 1980

Region, Country	Secondary Cities								Largest Cities					
	Number		Population (in thousands)		% of Urban Population		% Population Growth		Population (in thousands)		% of Urban Population		% Population Growth	
	1970	1980	1970	1980	1970	1980	1960-70	1970-80	1970	1980	1970	1980	1960-70	1970-80
East Africa														
Burundi	0	0	0	0	–	–	–	–	74	98	–	–	15.6	32.4
Ethiopia	1	1	226	439	9.7	9.6	21.0	94.2	784	1668	33.8	36.6	103.6	112.7
Kenya	1	1	256	396	22.3	17.8	58.0	54.6	550	1275	48.0	57.4	131.1	131.8
Madagascar	0	0	0	0	–	–	–	–	373	625	38.1	36.4	49.8	67.6
Malawi	0	0	0	0	–	–	–	–	148	352	36.4	18.8	–	137.8
Mauritius	0	0	00	0	–	–	–	–	136	153	39.3	30.2	22.5	12.5
Mozambique	0	0	0	0	–	–	–	–	375	750	80.1	83.2	106.0	100.0
Somalia	0	0	0	0	–	–	–	–	190	377	29.5	34.2	–	98.4
Uganda	0	0	0	0	–	–	–	–	357	813	45.6	51.5	134.9	155.7
Tanzania	0	0	0	0	–	–	–	–	375	1075	40.8	50.4	128.7	186.7
Zambia	5	5	732	1323	56.7	59.2	–	59.2	299	791	23.2	35.4	–	164.5
Middle Africa														
Angola	0	0	0	0	–	–	–	–	465	959	54.8	63.6	115.3	106.2
Central African Republic	0	0	0	0	–	–	–	–	187	297	37.2	36.3	55.8	58.8
Chad	0	0	0	0	–	–	–	–	155	313	37.4	39.3	–	101.9
Cameroon	1	1	178	352	15.0	14.4	130.6	97.8	250	526	21.1	21.5	44.5	110.4
Zaire	9	9	1916	3379	29.0	30.5	–	76.4	1367	3089	20.9	27.9	168.0	125.9

North Africa														
Algeria	6	6	1194	1682	18.3	13.9	53.6	40.8	1075	1391	16.5	11.5	23.1	29.4
Egypt	14	14	4577	5911	32.5	30.9	44.7	29.1	5480	7464	38.9	39.0	47.1	36.2
Libya	1	1	213	396	32.1	28.7	104.8	85.9	388	880	58.4	63.7	122.9	126.8
Morocco	9	9	2861	4292	54.6	51.9	42.6	50.0	1525	2194	29.1	26.6	38.5	43.9
Sudan	2	2	246	452	9.5	8.5	–	83.7	771	1621	29.9	30.6	11C.1	110.2
Tunisia	1	1	243	305	10.8	8.9	58.8	25.5	760	1046	34.C	30.8	2E.7	37.6
West and South Africa														
Benin	0	1	0	114	–	10.4	–	–	204	685	47.4	62.9	–	235.8
Botswana	0	0	0	0	–	–	–	–	–	110	–	47.0	–	–
Ghana	2	2	512	775	20.4	18.8	46.2	51.4	754	1416	30.C	34.5	9C.4	87.8
Guinea	1	1	0	174	–	8.2	–	–	330	736	60.8	79.8	182.1	131.2
Ivory Coast	0	0	0	0	–	–	–	–	356	685	29.8	32.6	97.8	92.4
Mali	0	0	0	0	–	–	–	–	249	440	33.2	34.2	70.8	76.7
Mauritania	0	0	0	0	–	–	–	–	–	198	–	38.9	–	–
Namibia	0	0	0	0	–	–	–	–	–	135	–	29.0	–	–
Niger	0	0	0	0	–	–	–	–	–	206	–	31.2	–	–
Nigeria	22	24	5049	9671	56.0	65.3	133.1	91.5	1389	2517	12.6	16.9	9.4	8.1
Senegal	0	2	0	218	–	17.2	–	–	559	821	60.1	64.9	59.3	46.9
Sierra Leone	0	1	0	145	–	17.4	–	–	202	388	42.5	46.5	96.1	83.0
Togo	0	0	0	0	–	–	–	–	150	273	58.4	60.4	–	82.0
Upper Volta	0	0	0	0	–	–	–	–	–	123	–	–	–	–

SOURCE: See source note, Table 2.1

NOTE: All 1980 estimates based on number and population of cities in 1970.

TABLE 2.3 Number, Population, and Growth Rates of Secondary and Largest Cities in Developing Countries of Central and South America: 1970 and 1980

Region/ Country	Secondary Cities								Largest Cities						
	Number		Population (in thousands)		% of Urban Population		% Population Growth		Population (in thousands)		% of Urban Population		% Population Growth	
	1970	1980	1970	1980	1970	1980	1960-70	1970-80	1970	1980	1970	1980	1960-70	1970-80
Central America														
Cuba	5	5	846	1200	16.4	17.4	144.7	40.9	1751	2139	33.9	31.0	20.9	22.2
Dominican Rp.	1	1	270	504	15.4	16.3	92.8	86.6	900	1661	51.4	53.8	93.9	84.6
Haiti	0	0	0	0					419	689	50.1	55.8	75.3	64.9
Jamaica	0	0	0	0					546	706	69.7	65.2	28.7	29.3
Costa Rica	0	0	0	0					452	637	65.6	64.2	48.2	40.9
El Salvador	0	1	0	124		0.6			336	433	24.3	21.9	32.3	28.9
Guatemala	0	0	0	0					733	1004	38.8	36.3	34.7	36.9
Honduras	0	0	0	0					235	423	32.0	33.1	80.8	80.0
Mexico	30	30	9019	15019	30.4	32.2	103.9	66.5	8997	15032	30.3	32.2	75.7	67.0
Nicaragua	0	0	0	0					411	683	44.2	46.9	65.7	66.2
Panama	0	0	0	0					443	695	63.7	66.3	64.0	56.8
South America														
Argentina	14	15	4669	6017	25.1	26.9	28.8	28.9	8469	10084	30.1	45.2	22.2	19.1
Chile	6	6	1028	1354	14.6	14.8	93.6	31.7	2889	3977	40.9	43.6	48.2	38.4
Uruguay	0	0	0	0					1312	1439	54.1	52.5	11.6	9.7
Bolivia	3	3	501	835	37.3	41.1	304.0	66.7	615	893	45.8	43.9	44.4	45.2
Brazil*	43	43	14995	26267	28.2	31.9	124.9	75.2	15101	24194	28.4	29.4	69.2	60.2
Colombia	23	23	7028	10492	53.2	49.5	92.3	49.3	2776	5493	21.0	25.8	112.0	97.8
Ecuador	1	2	526	967	22.1	26.0	59.4	83.8	730	1093	30.6	29.5	58.4	49.7
Guyana	0	0	0	0					165	193	100.0	100.0	11.4	16.9
Paraguay	0	0	0	0					379	529	44.4	43.9	35.4	39.6
Peru	8	8	1245	2374	16.3	19.8	386.3	90.7	2934	4682	38.6	39.2	66.9	59.5
Venezuela	10	10	2271	3682	28.2	31.2	146.8	62.1	2111	3093	26.2	26.3	58.1	46.5

SOURCE: See source note, Table 2.1.

*Rio de Janeiro and São Paulo are combined in largest cities category.

TABLE 2.4 Number, Population, and Growth Rates of Secondary and Largest Cities in the Middle East: 1970 and 1980

Country	Secondary Cities								Largest Cities					
	Number		Population (in thousands)		% of Urban Population		% Population Growth		Population (in thousands)		% of Urban Population		% Population Growth	
	1970	1980	1970	1980	1970	1980	1960-70	1970-80	1970	1980	1970	1980	1960-70	1970-80
Iran	19	24	4192	7855	35.6	40.9	103.1	90.2	3264	5447	28.1	28.4	71.3	66.9
Iraq	7	8	1608	3045	29.4	32.3	139.3	89.4	2510	5138	46.0	54.5	145.1	104.7
Jordan	1	2	140	351	12.3	19.7	—	150.7	394	655	34.8	36.6	75.1	66.2
Kuwait	1	2	109	396	18.7	31.1	—	263.3	224	404	38.5	31.7	43.8	80.4
Lebanon	1	1	183	240	11.9	9.4	32.6	31.1	1106	2003	72.4	78.6	108.3	81.1
Saudi Arabia*	4	5	891	1685	23.6	24.2	122.2	89.1	1096	2293	29.1	32.9	117.5	109.2
Syria	4	4	1106	1656	40.8	38.6	64.8	33.2	912	1406	33.7	32.8	56.4	54.2
Turkey	19	20	5839	9389	43.1	43.7	73.3	60.7	2760	5162	20.5	24.0	90.6	86.4
Yemen (North)	0	0	0	0	—	—	—	—	255	358	55.3	50.3	34.9	40.4
Yemen (South)	0	0	0	0	—	—	—	—	111	199	31.9	25.1	—	79.2

SOURCE: See source note, Table 2.1.

*Riyadh and Jeddah are combined in largest cities category.

TABLE 2.5 Number, Population, and Growth Rates of Secondary and Largest Cities in East and Southeast Asia: 1970 and 1980

Country	Secondary Cities								Largest Cities					
	Number		Population (in thousands)		% of Urban Population		% Population Growth		Population (in thousands)		% of Urban Population		% Population Growth	
	1970	1980	1970	1980	1970	1980	1960-70	1970-80	1970	1980	1970	1980	1960-70	1970-80
Bangladesh	5	5	1532	2856	29.7	29.9	31.0	86.4	1289	2841	25.0	29.8	147.9	120.4
Burma	5	5	1029	1536	16.2	16.1	63.5	49.3	1453	2185	22.9	22.8	48.5	50.4
China, P.R.[a]	137	–	91615	–	51.8	–	56.4	–	16500	–	9.3	–	29.9	–
India**	135	144	45482	68958	42.5	44.6	59.8	51.6	12722	17165	11.9	11.1	33.1	34.9
Indonesia	26	26	8980	12531	44.0	40.0	43.1	39.5	4450	7263	21.8	23.2	64.3	63.2
Korea, North	8	8	1940	3171	27.8	29.6	118.2	63.4	911	1283	13.1	11.9	43.5	40.8
Korea, South	17	17	6130	10152	48.0	40.6	91.3	65.6	5322	8490	41.6	40.5	125.3	59.5
Malaysia	4	4	770	1100	27.2	26.7	99.5	42.9	649	1106	22.9	26.9	75.4	70.4
Nepal	0	0	0	0	–	–	–	–	146	190	33.2	26.8	23.7	30.1
Pakistan	15	20	6077	10025	40.4	42.8	49.4	64.9	3139	5005	20.9	21.4	56.9	59.4
Philippines	15	15	2629	4153	21.2	21.9	127.8	57.9	3591	5664	28.9	29.9	56.9	57.7
Sri Lanka	2	2	258	337	9.4	8.2	155.4	15.3	561	647	20.5	15.7	14.9	15.3
Taiwan[a]	11	–	3270	–	35.2	–	76.5	–	2150	–	23.0	–	49.0	–
Thailand[b]	1	2	111	237	–	3.4	–	113.5	3205	4870	67.8	68.5	49.0	51.9

SOURCE: See source note, Table 2.1.

a. 1970 data from Kingsley Davis, *World Urbanization 1950-1970* (Berkeley: University of California, 1969). United Nations reports combine the Republic of China and the People's Republic of China for 1980 estimates.

b. 1975 data from Government of Thailand, Economic and Social Development Board, 1976.

*Peking and Shanghai are combined in largest cities category.

**Bombay and Calcutta are combined in largest cities category.

In much of the rest of the developing world, however, the population of all secondary cities increased more slowly than that of the largest city. In Zambia, Zaire, Algeria, Egypt, Morocco, and Nigeria, the populations of the largest cities grew by an average of 88 percent between 1970 and 1980, compared to an average of 65 percent in secondary urban centers. In only 4 African countries did the combined population of secondary cities grow faster than that of the largest city during the 1970s. In half of the 16 countries in Africa with at least 1 city over 100,000 population, the largest city grew at a rate much higher than the average in secondary cities during both the 1960s and 1970s (see Table 2.2).

In Central America, only Cuba and Mexico had more than one secondary urban center and in neither country did the number increase. The population growth in all secondary cities of Central American countries between 1970 and 1980 was less than three-quarters of the population increase in the largest cities. In the Middle East, where secondary cities have been growing rapidly in number and population, the largest cities had higher population growth rates during the 1970s (see Tables 2.3 and 2.4).

United Nations demographic studies indicate that cities of any size that were national capitals, and the largest cities, grew more rapidly than noncapital cities. In many developing countries even smaller cities that were capitals grew faster than the largest and secondary cities that were not. "Being the largest city in a country confers a growth increment about half as large as that pertaining to capitals," analysts note. Although the small number of cities involved made the results statistically insignificant, the analysts point out that the trends "are clearly consistent with the view that spatial patterns of government expenditure bias patterns of city growth toward the largest city and particularly toward capital cities."[6]

Secondary cities have played a relatively weak role in absorbing population increases in most developing countries or in creating a more balanced spatial distribution of population. The relatively weak role that secondary cities play in absorbing population growth or balancing the spatial distribution of population is seen clearly during the period 1950 to 1970. Even though the number and population of secondary cities increased significantly, their relative share of population did not keep pace with that of the largest cities or of urban places under 100,000. Davis points out that there was a 15 percent shift in population from rural to urban areas in Latin America between 1950 and 1970, yet cities in the 100,000 to 500,000 range increased their share of population by less than 1 percent during the same period and cities of from 500,000 to 1 million increased their share by only a little more than 2 percent. A great number of rural migrants went to the largest cities in South America.[7] United Nations estimates for 1980 indicate that secondary cities account for one-third or more of the urban population only in Bolivia and Colombia. In Argentina, Chile, Bolivia, Ecuador, and Peru, the population of the largest city is substantially higher than the combined population of all secondary urban centers in the country. In Brazil, the population of 43 secondary cities is only slightly larger than that of Rio de Janeiro and São Paulo. Nearly 40 percent of the urban population in South American countries remained in urban places smaller than 100,000 and about 43 percent lived in the largest city.

Similarly, in Central America, the largest cities in all countries except Mexico were under 1 million in population in 1970, but they increased their share of population by nearly 10 percent since 1950, while cities of between 100,000 and 500,000 increased their portion by only about 3.5 percent. Population estimates for 1980 show that only 3 of the 11 countries had any cities over 100,000 population and in all 3 the largest city's population exceeded that of secondary cities. Only in Mexico did secondary cities account for more than one-third of the urban population. About 45 percent of the urban population in Central America was concentrated in the largest cities in 1975 and about 42 percent of urban dwellers were living in places smaller than 100,000. Fox and Huguet point out that cities of between 100,000 and 250,000 population in Central America have been *declining* in number and percentage of urban population since the 1950s. Most of the cities larger than 250,000 are national capitals, and they have been growing rapidly. Fox and Huguet observe that "this underscores the very weak position of secondary cities throughout most of the region, and throughout Latin America for that matter. Very few centers can compete outside the capitals in growth and prosperity."[8]

Similar conclusions can be drawn about Asia, which has the largest number of secondary cities and the most population living in them of any developing region. In East Asia less than one-fifth, and in South Asia less than one-third, of the urban population was found in secondary cities in 1975. Nearly 40 percent of Asia's urban dwellers live in places of less than 100,000 population. The weak role of Asian secondary cities is seen in the distribution of urban population: In 1950 cities of from 100,000 to 500,000 had only 2.8 percent of the population and increased their share by less than 1 percent by 1970; cities of from 500,000 to 1 million gained less than one-half of 1 percent; and in cities of from 500,000 to 1 million the share of population declined. In Indonesia, the total population living in cities of more than 1 million increased from about 4.1 to 6.2 percent during the 1960s, while the share of population living in cities of from 100,000 to 1 million declined slightly. Hugo notes that in Java only the 2 largest cities, Jakarta and Surabaya, grew at high rates and that Indonesian cities of less than 500,000 grew at a rate lower than that of national population growth, pointing "to a general stagnation in the middle levels of Java's urban hierarchy."[9]

In Asia, as in much of the developing world, migration from rural areas and small towns has played a major role in the expansion of primate cities and in the growth of the largest metropolitan areas. In Thailand and the Philippines, the most frequent destination of rural migrants is the primate city: In Thailand rates of inmigration and growth vary positively, and outmigration inversely, with city size; in the Philippines, Manila has attracted nearly half of all rural migrants to cities over the past 20 years. The World Bank estimates that during the 1950s half of Bombay's growth could be attributed to migration, as could 60 percent of Jakarta's during the 1960s. About 63 percent of Seoul's growth from 1955 to 1965, and 43 percent of Taipei's from 1960 to 1967, can be attributed to migration. The relatively weak role that secondary cities play in absorbing population growth, even in India, is reflected in the fact that cities of over 100,000 accounted for less than 10 percent of India's total population in 1971. Less than 3 percent of Indians lived in cities of from 50,000 to 100,000; more than 80 percent were scattered in a half million villages containing less than 10,000 people each and more than half of all settlements had less than 500 people.[10]

Secondary cities seem to be weakest in Africa and the Middle East. Urban population increased by more than 11 percent in the Middle East between 1950 and 1970, but secondary urban centers absorbed less than 2.5 percent.[11] Less than 30 percent of the urban population in the Middle East now lives in secondary cities. Less than half of the African developing countries have even 1 city of this size, and in 12 of these 17 countries the capitals have a larger population. Only in Nigeria, Zambia, and Morocco do secondary cities have a large percentage of the urban population. Cairo's population is estimated at nearly 3 million larger than the combined populations of Egypt's 14 secondary cities, which have an estimated 31 percent of the urban population—about 1.5 percent *less* than in 1970. There were no cities in the 500,000 to 1 million population range in 1975, and cities of from 100,000 to 500,000 account for a little more than 18 percent of Egypt's urban population. El-Shakhs points out that "the development of large regional urban centers to fill the gap between the two primate cities and the rest of the urban system has been conspicuously lagging behind national urbanization and industrialization trends." He notes that the share of urban population in Egypt's secondary cities has declined.[12]

Economic and Social Characteristics

Reliable data on the social and economic characteristics of secondary cities in developing countries are also rare; generalizations must be based on fragmentary census surveys and studies of individual cities. The specific functions that secondary cities perform will be reviewed in Chapters 3 and 4, but some broad generalizations about their socioeconomic characteristics are offered here.

Secondary cities tend to have a combination of urban and rural socioeconomic characteristics and they generally perform functions found in both urban areas and the countryside. One of the strongest impressions that evolves from a review of secondary cities in the developing world is that they have a blend of urban and rural characteristics. They are "middle cities" in many senses of the term: They share some social, economic, and physical characteristics with both larger metropolitan centers and smaller towns and villages. They are like the major metropolitan centers in that their economies are dominated by commerce and services, and some have large manufacturing sectors. Yet they often have a small share of the nation's manufacturing establishments and employment, and their industrial sector is usually composed of very small firms. Most secondary cities are not competitive with large metropolitan centers for industries, and even their commercial and service establishments are usually of small size.

Small secondary cities tend to have larger proportions of their labor force engaged in agriculture, agroprocessing, marketing, and farm services, making them dependent on their rural hinterlands' agricultural production, which is often low. Living conditions for many residents are better than in rural areas and small towns, yet are far below those found in the largest metropolitan areas.

Ulack notes in his study of Iligan City in the Philippines that as industrialization proceeds, the socioeconomic characteristics of a secondary city become increasingly different from those of smaller towns. He points out that secondary centers have higher rates of inmigration, and younger, relatively better educated, and more culturally heterogeneous people. Fertility rates among women begin to decline and the middle income group becomes larger.[13] Yet even as these changes occur, secondary centers remain a blend of urban and rural lifestyles. Hazelhurst observed that in India "their multiformity is typically reflected in a patchwork of small-scale industrial and commercial enterprises, for they incorporate economic functions of both market towns and small-scale manufacturing centers."[14] He vividly describes the blending of city and country, observing that

to the occasional visitor, middle range cities contain a bewildering variety of economic activity. Often the first glimmering of the city's economic life is encountered at the railroad station or bus stand, typically congested with tea stalls, sweet shops and ricksha drivers. Turning to the streets of the bazaar, one finds some areas heavily populated with hawkers, who might be selling fruit and vegetables or odds and ends of cloth. . . . Behind the hawkers are the shops of resident businessmen. One finds, for example, bicycle shops and bicycle mechanics, small-scale metal workers who manufacture trunks and agricultural implements, the shops of medical practitioners and dispensers of medicine, cloth shops and brass shops, tailors and dry cleaners, radio repair shops and shops labeled simply "general merchandise," gold merchants and grain merchants. . . . Both traditional consumer goods and luxury goods of more recent origin can be found in the bazaar of the middle range city.[15]

The mingling of urban and rural activities persists even as secondary cities grow to a half million or more people. Periodic markets still play an important role in the economy of Ranchi, India, a regional administrative and industrial center in Bihar. With more than a half million residents, it serves as an important consumer and agricultural market for its own population and for its rural hinterland. Srivistaba notes that the mix of urban and rural activities is especially visible in the periodic markets, which are particularly large in Greater Ranchi, drawing buyers from urban and rural areas and sellers from distant rural villages. He describes the bi-weekly markets of Ranchi as being "as large as to resemble a small fair." They bring together urban merchants and rural farmers to exchange vegetables, rice, grains, fruit, goats, agricultural implements, woven cloth, manufactured goods, and a variety of other products.[16]

The economies of secondary cities tend to be dominated by commercial and service activities, with manufacturing employment concentrated primarily in the small-scale industrial sector. Secondary cities tend to differ in their economic composition and the mix of activities that generate income and employment, depending on size, location, and patterns of previous investment. Cities of different sizes provide different advantages to economic activities. Lo and Salih's study of Asian countries concludes that the following occupational structures are characteristic of cities in different size classifications:[17]

(a) Cities with populations smaller than 100,000 have high proportions of employment in agriculture and related marketing and commercial activities, in small-scale cottage and artisan manufacturing, and lower-order services that have a relatively low growth rate in total urban employment.

(b) Cities with a population of between 100,000 and 250,000 have generally high rates of employment in small-scale manufacturing and in consumer-oriented commercial and service activities and have relatively high rates of total urban employment.

(c) Cities with populations of from 250,000 to one-half million are characterized by an increasing rate of growth in the producer-oriented commercial sector. They tend to have substantial manufacturing and tertiary activities, with increasing rates of growth in the producer-oriented commercial and services sectors.

(d) Cities of one million or more have a relatively high proportion of employment in manufacturing, but their occupational structure is dominated by producer-oriented commercial and service sectors.

Variations in the economic structures of cities in different size groups may be explained in part by their economies of scale. Cities smaller than 100,000 may not have sufficient population to support large-scale commercial and manufacturing activities that are dependent on local markets. They do offer sufficiently large markets, however, to sustain lower-order, small-scale consumer and service functions, and agricultural marketing and service activities. As cities increase in size they begin to offer economies of scale and proximity that allow larger volumes of production and generate demand for producer-oriented commercial goods and services, and thus allow secondary and tertiary sectors to operate more efficiently. Lo and Salih suggest that in Asia manufacturing efficiency increases with city size up to about 1 million and then begins to decline, while producer-oriented tertiary activities continue to increase in efficiency with population size after cities pass the 1 million mark.

Although evidence to validate these hypotheses—or any others, for that matter—is fragmentary, what exists seems to support them. South Korea offers an example of the changes that seem to occur in urban structure with continuing urbanization, rapid industrialization, and deliberate attempts by government to deconcentrate economic activities from the primate city, all of which happened in Korea between 1960 and 1980. At the beginning of the 1960s, South Korea was among the poorest developing countries, just recovering from a devastating war in which much of its manufacturing capacity and agricultural resources had been destroyed. Two-thirds of the labor force was engaged in agriculture and living in rural areas. In 1960, Korea's urban structure was not much different from that of many contemporary developing nations. Other than Seoul, only 8 cities had populations larger than 100,000. Only Pusan and Taegu had more than a half million residents. The national capital clearly dominated both the economy and the spatial system of the country. With few exceptions, nearly all of the cities that had grown to 30,000 or more in population were commercial, service, or agricultural market centers.

Agricultural employment was still relatively strong in many smaller cities in Korea in 1960. About 23 percent of the urban labor force was engaged in primary sector employment. Although on average only about 14 percent of the labor force in cities with populations over 90,000 worked in primary activities, nearly one-third of the workers in cities of from 50,000 to 90,000 people were so employed. Of the 13 cities with less than 90,000 people, 7 were relatively specialized in agriculture, but only 1—Jeju, a remote island provincial capital—had a share equal to or greater than that of the national economy (see Table 2.6). Moreover, other than Seoul, only 4 cities—and none less than 90,000 in population—had more than 20 percent of their labor force employed in manufacturing. Nearly all cities with populations greater than 90,000 had some manufacturing capacity and all but Gwangju—a provincial capital in the agricultural southwestern region—had a larger percentage of their labor force in manufacturing than that of the urban sector and the national economy.

Clearly, however, every city that had grown to 50,000 or more people in 1960 had heavy concentrations of employ-

TABLE 2.6 Percentage Distribution of Employment and Location Quotients of Employment in Secondary Cities of Korea: 1960

City	Population 1960 (in thousands)	Agriculture % of Labor Force	Agriculture L.Q.	Agriculture Urban National L.Q.	Manufacturing % of Labor Force	Manufacturing L.Q.	Manufacturing Urban National L.Q.	Commerce % of Labor Force	Commerce L.Q.	Commerce Urban National L.Q.	Services % of Labor Force	Services L.Q.	Services Urban National L.Q.	% of Labor Force In Tertiary Sector
Pusan	1,163.7	4.9	.21	.07	22.8	*1.42	*3.35	22.8	*1.21	*1.44	34.5	.84	*4.11	57.3
Taegu	676.7	13.6	.59	.21	29.4	*1.83	*4.23	20.6	*1.09	*1.30	26.9	.65	*3.20	47.5
Incheon	401.5	14.2	.62	.21	18.7	*1.16	*2.75	17.1	.90	*1.08	33.7	.82	*4.01	50.8
Gwangju	314.4	33.2	*1.45	.50	14.1	.88	*2.07	15.8	.84	1.00	28.9	.70	*3.44	44.7
Daejeon	228.9	5.4	.24	.08	21.6	*1.35	*3.17	23.4	*1.24	*1.48	36.1	.88	*4.30	59.5
Jeonju	188.2	25.7	*1.12	.40	17.3	*1.08	*2.54	15.0	.80	.94	33.6	.82	*4.00	48.6
Masan	158.0	7.9	.34	.12	18.4	*1.15	*2.70	23.5	*1.24	*1.49	38.6	.94	*4.59	62.1
Mogpo	129.7	7.4	.32	.11	16.9	*1.05	*2.49	26.1	*1.38	*1.65	33.7	.82	*4.01	59.8
Cheongju	92.1	9.4	.41	.14	19.3	*1.20	*2.83	22.5	*1.19	*1.42	38.5	.94	*4.58	61.0
Suweon	90.8	9.2	.40	.14	19.8	*1.24	*2.91	21.7	*1.15	*1.37	34.7	.84	*4.13	56.4
Gunsan	90.4	12.3	.53	.19	22.2	*1.39	*3.26	31.4	*1.13	*1.35	32.2	.79	*2.83	53.6
Yeosu	87.2	20.6	.90	.31	9.9	.62	*1.45	23.6	*1.25	*1.49	33.5	.82	*3.99	57.1
Jinju	87.1	25.4	*1.11	.38	18.7	*1.17	*2.75	16.3	.86	*1.03	30.1	.74	*3.58	46.4
Chuncheon	82.5	10.7	.47	.16	11.3	.70	*1.66	20.9	*1.10	*1.32	44.0	*1.08	*5.23	64.9
Weonju	76.9	14.1	.62	.21	9.9	.62	*1.45	20.6	*1.09	*1.30	42.3	*1.03	*5.03	62.9
Gyeongju	75.9	51.5	*2.25	.78	8.3	.52	*1.22	14.9	.79	.94	18.3	.45	*2.17	33.2
Suncheon	69.5	51.7	*2.25	.78	7.4	.46	*1.09	12.0	.63	.76	19.9	.49	*2.36	31.9
Chungju	68.7	32.2	*1.41	.49	15.3	.96	*2.25	17.4	.92	*1.10	23.4	.57	*2.79	40.8
Jeju	67.9	67.0	*2.92	*1.01	4.8	.30	.71	8.2	.43	.51	14.1	.34	*1.68	22.3
Jinhae	67.7	11.7	.51	.18	7.1	.45	*1.04	12.9	.68	.82	58.8	*1.44	*7.00	71.7

Iri	65.8	17.4	.76	.26	16.9	*1.06	*2.48	19.9	*1.05	*1.26	34.9	.85	*4.15	54.8
Pohang	59.5	23.8	*1.04	.36	12.1	.75	*1.78	21.0	*1.11	*1.33	30.6	.75	*3.64	51.6
Gangneung	58.7	40.3	*1.76	.61	11.3	.71	*1.57	14.5	.77	.92	22.7	.56	*2.70	37.2
Andong	53.4	18.5	.81	.28	12.8	.80	*1.89	22.3	*1.18	*1.41	33.6	.82	*4.00	55.9
Urban Sector		22.9			16.0			18.9			40.9			59.8
Nation		66.4			6.8			15.8			8.4			24.2

SOURCE: Calculated from Republic of Korea, [Long Range Planning of Urban Growth to the Year 2000: Data Collection], vol. 1 (unofficial trans.) (Seoul: Ministry of Construction, 1980), Table I/2-7.

a. Location quotient for employment in urban areas:

$$\text{Urban L.Q.} = \frac{e_i/e_t}{E_i/E_t}$$

where:

e_i = number of workers employed in sector i in city
e_t = total number of workers employed in city
E_i = number of workers employed in sector i in all urban areas
E_t = total number of workers employed in all urban areas

b. Location quotient for employment in national economy:

$$\text{National L.Q.} = \frac{e_i/e_t}{N_i/N_t}$$

where:

e_i = number of workers employed in sector i in city
e_t = total number of workers employed in city
N_i = number of workers employed in sector in nation
N_t = total number of workers employed in nation

*Relatively specialized compared to base area.

ment in the tertiary sector. Every city larger than 90,000 had at least 44 percent of its labor force engaged in commerce and services, and all except 1 with from 50,000 to 90,000 people had at least 30 percent of their labor force working in tertiary activities. Of the cities with more than 90,000 residents, about three-fourths had at least half of their labor force working in commerce and services (see Table 2.6). The concentration is reflected in service employment quotients. All urban settlements had about 41 percent of their workers employed in services, compared to 8 percent of the national labor force, and cities of over 50,000 residents had from 2 to 7 times the number of service workers of the national economy.

In brief, those cities with more than 90,000 in population —the secondary cities of Korea in 1960—were places in which agricultural employment was relatively weak. Manufacturing employment was highly concentrated in Seoul and some large cities and only a few secondary cities had more than one-fifth of their labor force working in industry. All of the secondary cities in Korea in 1960 were primarily commercial and service centers, with more than 40 percent of their labor force engaged in tertiary activities; smaller urban centers remained either agricultural market and supply centers or had high concentrations of their workers in services and commerce as well.

During the 20-year period between 1960 and 1980, the Korean government concentrated its investments in export production, light and heavy industry, and infrastructure, while attempting to increase agricultural output and prices and improve rural living conditions to generate equitable growth. Deliberate efforts were made during the late 1960s and the 1970s to deconcentrate manufacturing and commercial employment from the Seoul metropolitan area, which had grown rapidly in population despite policies to reduce its growth. Yet, during the period, both the urban structure in Korea and the economies of secondary cities changed drastically and in ways that were compatible with the theories described earlier.

By 1980, the number of cities with populations of 100,000 or more increased to 30, and their economic structures changed drastically. In only 3 of these cities did the primary sector play a significant role in employment (see Table 2.7). Manufacturing employment had risen to an average of 55 percent in cities of 200,000 or more residents, and to 40 percent of the labor force in cities of from 100,000 to 200,000 residents. The most drastic changes, however, came in the tertiary sector: Employment in services in cities of 200,000 or more population dropped from over 43 percent to about 22 percent, and in cities of between 100,000 and 200,000 it dropped from an average of nearly 40 percent to a little more than 26 percent. Although the tertiary sector remained strong in the occupational structure of secondary cities, in those with more than 500,000 population, production-oriented services accounted for about 12 percent of employment, and personal services dropped to less than 10 percent.

By 1980, secondary cities in Korea began to exhibit a strong division of labor and a high degree of functional specialization. Employment location quotients indicate that 4 cities with less than 200,000 people and 1 with about 210,000 population became relatively specialized in agricultural processing, distribution, and related activities. Mogpo and Cheonan had about a 10 times greater share of their labor forces in primary sector activities than did all urban places in Korea, and Jinhae had about 5 times the agricultural work force of other urban places. Anyang and Gunsan had slightly stronger concentrations of agricultural workers than the average in urban places (see Table 2.8). A total of 9 cities with more than 160,000 population attained relative specializations in manufacturing. There were 4 cities with more than

TABLE 2.7 Changes in Labor Force Distribution Among Sectors in Secondary Cities of Korea, by Sector and Size Categories: 1960-1980

Population Size Category 1978	Number of Cities	Average % Employment in							
		Agriculture		Manufacturing		Commerce		Services	
		1960	1980	1960	1980	1960	1980	1960	1980
500,000 or More	5	14.5	1.4	21.3	55.6	19.9	21.7	43.2	22.2
499,999-200,000	7	17.4	6.6	16.9	55.2	17.4	16.5	44.4	21.2
199,999-100,000	18	28.1	4.7	14.1	39.7	17.5	29.2	39.8	26.3

SOURCES: Compiled from Republic of Korea, [Long Range Planning of Urban Growth to the Year 2000: Data Collection], vols. 1, 2 (unofficial trans.) (Seoul: Ministry of Construction, 1980); and Republic of Korea, *Municipal Yearbook of Korea, 1980* (Seoul: Mininstry of Home affairs, 1980).

300,000 people—Taegu, Gwangju, Ulsan, and Cheongju—that emerged as regional centers of production-oriented services, and also had relatively large numbers of workers in industry. Daejeon and Gangneung—smaller secondary cities—also emerged as strong centers of production-oriented services. All of the secondary cities with populations smaller than 150,000 remained highly specialized in commerce and services, and none attained relative specializations in manufacturing, although in 4 cities with between 150,000 and 200,000 residents, employment in industry had increased.

At least 7 functional types of secondary cities could be found in Korea by 1980. Mogpo, Cheonan, and Jinhae—3 smaller secondary cities—became agricultural processing and distribution centers. Anyang and Gunsan, with between 150,000 and 200,000 residents, had relatively high concentrations of workers in both agriculture and manufacturing. Pusan, Inchon, Masan, and Seongnam—all large secondary cities—and one smaller port city, Pohang, where the government concentrated investments in industry and industrial estates, clearly became manufacturing centers. Jeonju, also a larger secondary city, had relative specializations in both manufacturing and services. Andong, Iri, and Yeosu—smaller secondary cities—had relatively high concentrations of their labor force in commerce. Daejeon, Gwangju, Ulsan, Gangneung, Jinju, Bucheon, and Chungju had become service centers, and 9 other cities, 6 of which had less than 200,000 people, remained service-commercial centers.

Thus over a 20-year period both the urban structure of Korea and the occupational composition of secondary city economies changed markedly. Manufacturing employment increased in the larger secondary cities, agriculture disappeared as a substantial source of employment in all but a few, and the importance of services decreased in nearly all of them. Yet the tertiary sector remained an important part of the economies of all urban centers, and especially of those with less than 200,000 people.

TABLE 2.8 Distribution and Location Quotients of Employment for Secondary Cities of Korea: 1980

| City | Population 1978 (in thousands) | Employment Sector | | | | | | | | | |
| | | Agriculture | | Manufacturing | | Production-Oriented Services[a] | | Commerce | | Other Services | |
		% of Labor Force	Urban L.Q.	% of Labor Force	Urban L.Q.	% of Labor Force	Urban L.Q.	% of Labor Force	Urban L.Q.	% of Labor Force	Urban L.Q.
Pusan	2,879.6	.4	.10	70.1	*1.21	5.9	.81	13.4	.68	10.1	.92
Taegu	1,487.1	—	—	47.9	.82	7.5	*1.04	30.2	*1.52	14.3	*1.30
Incheon	963.5	.7	.18	72.2	*1.24	6.7	.93	17.9	.90	2.4	.21
Gwangju	694.6	—	—	32.8	.56	18.5	*2.57	32.2	*1.63	16.3	*1.48
Daejeon	580.6	.3	.08	55.3	.95	24.7	*3.43	14.7	.74	4.9	.44
Masan	391.9	.6	.15	77.4	*1.33	2.3	.32	12.5	.63	7.2	.65
Jeonju	384.1	—	—	69.4	*1.19	4.6	.63	4.9	.25	18.9	*1.72
Ulsan	364.5	—	—	40.2	.69	20.4	*2.83	25.7	*1.29	13.6	*1.23
Seongnam	324.1	—	—	76.8	*1.32	2.3	.32	16.3	.82	4.4	.40
Suweon	266.1	.2	.05	56.4	.97	5.9	.82	23.9	*1.20	13.5	*1.22
Cheongju	223.1	—	—	40.0	.68	17.9	*2.49	12.0	.60	29.7	*2.70
Mogpo	210.9	40.0	*10.26	26.1	.45	7.9	*1.10	20.5	*1.04	5.4	.49
Anyang	187.9	4.3	*1.10	77.2	*1.33	2.5	.35	10.0	.50	5.2	.47

Pohang	184.0	—	—	68.2	*1.48	6.1	.84	19.1	.96	6.6	.60
Jinju	174.9	—	—	30.3	.30	6.5	.90	22.4	*1.13	40.8	*3.70
Gunsan	167.4	5.9	*1.15	76.6	.24	2.6	.36	10.4	.52	4.3	.39
Bucheon	163.5	1.4	.35	85.8	.60	1.9	.26	8.9	.45	1.9	.17
Chuncheon	152.6	—	—	18.0	.36	8.2	*1.38	25.6	*1.29	48.0	*4.36
Jeju	152.5	3.7	.95	14.3	.61	13.7	*1.90	41.0	*2.07	27.3	*2.48
Yeosu	151.3	1.7	.43	34.7	.71	8.4	*1.17	46.6	*2.35	8.5	.77
Iri	132.3	.3	.07	20.9	.42	8.9	*1.24	55.3	*2.79	14.5	*1.32
Weonju	131.0	.6	.15	35.4	.61	4.8	.66	39.2	*1.97	19.9	*1.80
Euijeongbu	117.8	—	—	41.0	.71	10.6	*1.47	33.4	*1.68	14.8	*1.34
Suncheon	114.6	2.0	.51	24.9	.42	14.3	*1.98	41.6	*2.10	17.0	*1.54
Gyeorgju	113.9	—	—	35.4	.61	6.0	.83	39.7	*2.00	18.8	*1.71
Chungju	110.1	1.5	.38	39.0	.67	7.4	*1.02	25.8	*1.30	26.2	*2.38
Cheonan	109.3	42.3	*10.84	40.9	.70	4.8	.66	8.3	.42	3.6	.33
Jinhae	108.7	20.2	5.78	30.3	.52	7.3	*1.01	28.3	*1.42	13.7	*1.24
Gangneung	102.2	.2	.05	25.5	.44	41.3	*5.74	6.1	.31	26.8	*2.43
Andong	101.5	—	—	16.5	.28	10.5	*1.45	63.0	*3.18	9.8	.89
Urban areas		3.9	—	58.1		7.2		19.8		11.0	

SOURCE: Calculated from Republic of Korea, *Municipal Yearbook of Korea, 1980* (Seoul: Ministry of Home Affairs, 1980), Table 36.

a. Includes gas, utilities, construction, transportation, warehousing, and communications.

*Relatively specialized compared to all urban areas.

In countries that have not experienced Korea's rates of urbanization and industrialization, the characteristics of secondary cities remain much like those in Korea during the 1950s and 1960s. Manufacturing employment remains concentrated in the primate city, the few secondary cities that have attained significant levels of nonagricultural employment have economies that are dominated by the tertiary sector, and the smaller cities and towns are still primarily agricultural and rural service centers. This pattern is quite clearly seen in Thailand, where 33 percent of the labor force was engaged in manufacturing in Bangkok in the early 1970s and nearly 60 percent in commerce and services. In the two cities with more than 100,000 population, Chiangmai and Songkhla, only a very small percentage of the economically active population was engaged in manufacturing—nearly 77 percent of employment in Chiangmai was in agriculture, as was about 80 percent in Songkhla.[18]

The primary differences in the economic structures of Manila and intermediate cities in the Philippines are reflected in occupational profiles of the labor force in the primate city in 1970 and in Davao City in 1972. Although both had less than 4 percent of their work force engaged in agriculture and about 40 percent in services, significant differences appeared in the size of the labor force in manufacturing and commerce. Nearly 29 percent of Manila's labor force was employed in manufacturing, compared to less than 19 percent of Davao's. The Philippines' second largest city had nearly one-fourth of its labor force employed in commercial and sales activities, while only 14 percent of Manila's workers were similarly employed. Not only is a far larger percentage of the secondary city's workers employed in tertiary sectors, but half of Davao's commercial employment was in "informal" bazaar activities—vending, peddling, and small-scale trading. Hackenberg's studies of Davao City indicate that by 1978 the percentage of employment in manufacturing had declined from 19 to less than 12. Employment in commerce increased to about 30 percent, with the increase in the share of the labor force earning a living in the bazaar sector accounting for nearly all of the change.[19]

The economic base of Philippine secondary cities is dominated by the informal tertiary sector, which is composed of very small commercial activities in which most participants barely manage to eke out a living. Hackenberg notes that the majority of the labor force in Davao City is engaged in "preindustrial" activities. Moreover, this informal commercial sector is seen as the major channel of upward mobility for most of the poorer residents of secondary cities because of its low level of skill and educational requirements, the ability to increase income by adding more family members, and the low costs of operation.[20] Hackenberg's surveys of Davao City in the mid-1970s found that "the portion of the labor force participants directly involved in formally organized manufacturing was not of sufficient size (only 11.3% of those employed) to play a major role in shaping the contours of urban society."[21] The manufacturing sector consisted of 48 firms with 10 or more workers and 759 small establishments employing 9 people or less. The 12 industries with 100 or more workers employed two-fifths of the 11 percent of the labor force engaged in manufacturing, indicating that the far larger small-industry sector could absorb relatively little labor.

As will be seen in Chapter 4, the manufacturing and commercial sectors in secondary cities tend to be composed of large numbers of small industries and shops, often family owned and employing few workers. Secondary cities in many developing nations have little or no large-scale industry.

Secondary cities tend to have a relatively small share of manufacturing activities and employment, compared to their share of population, than the largest cities, and they contrib-

ute less to national output. What is most striking about the economies of secondary cities in most developing countries is that they have relatively small shares, compared to the largest city, of manufacturing activity and employment. This has been the case historically in colonized countries because it was the policy of most colonial governments, and post-independence regimes as well, to concentrate investments in infrastructure and productive activities in the largest cities, and especially in the national capital. Small towns and secondary cites usually received little industrial invest-ment.[22] Thus in Kenya, industrial activities are highly con-centrated in Nairobi, and relatively small shares of industrial employment are found in other cities. In 1975, nearly 57 percent of manufacturing employment was in Nairobi, while among the next 3 largest cities Mombasa had only 17 per-cent, Nakuru less than 6 percent, and Kisumu a little more than 3 percent.[23] Indeed, 58 percent of the country's total wage employment was accounted for by Nairobi, with Mombasa, Nakuru, and Kisumu having only 19.0, 4.3, and 4.9 percent, respectively.

Analysts also note the heavy concentration of Egypt's industrial activities in Cairo and Alexandria, and the weak-ness of the industrial base in secondary cities. Only Alex-andria, they observe, "has managed to hold out against Cairo's magnetism. Sixty-five percent of the economically active labor force is concentrated in the two cities. Thus it is not surprising that 50 percent of industrial enterprises in Egypt are located within their boundaries of their immediate hinterlands."[24]

Even in relatively industrialized developing nations, such as Taiwan, large manufacturing firms tend to remain concen-trated in the largest city. The 3 secondary cities of Taichung, Tainan, and Kaoshiung combined had less than half the number of firms with 10 to 500 workers than did Taipei during the 1970s. These 3 cities had about 20 percent of all manufacturing establishments with more than 500 workers, while Taipei alone accounted for one-third of these industries. The 3 secondary cities contained a larger percentage of small-scale industries employing less than 10 workers, and these small plants formed the base of the intermediate cities' manufacturing sector.[25]

Secondary cities tend to have a greater diversity and better quality of social services and facilities than smaller towns and rural villages, but have a proportionately smaller share and poorer quality of services and facilities than the largest cities. The intermediacy of secondary urban centers is also reflected in the quantity and quality of their social services and facilities. Atmodirono and Osborn observe that Surabaya, Bandung, Semarang, Medan, Palembang, and Udung Pandang range from one-half to one-eighth the size of Indonesia's capital city, Jakarta, "but they are economically inferior by much greater margins in terms of investment, port and manufacturing functions." They also point out that these cities rank far behind Jakarta in commerce and services, but ahead of smaller towns and rural areas: "Common deficiencies in services are low levels of welfare services as related to population, though water, electricity, roads, education and health facilities are generally higher in these middle cities per capita than elsewhere in their provinces."[26]

Analysis of urban services and facilities in Colombia concluded that the quantity and quality of educational, health, recreational, and other services in secondary cities ranked far below those of Bogotá during the 1960s and 1970s. Large gaps also appeared between the larger secondary cities and those with under one million population. The USAID Mission in Colombia observed that "the lack of adequate educational and recreational facilities (and health care, as well, if the low proportion of doctors is taken into account) in intermediate cities, are not only areas in which they are proportionately worse off than the four major cities. The same is true for housing and water and sewerage services."[27] Gilbert found in a study of educational and health services in Colombian cities and towns during the mid-1970s that "a strong positive relationship existed between service provision and size of city among the larger cities of the country." Moreover, he observed from empirical studies that "among these cities the quality of health and education provided in the larger cities was far superior to that in the smaller cities. There were no exceptions to this rule." The numbers of doctors and dentists per 1,000 population dropped steadily with declining population size, and dropped dramatically for cities with less than 100,000 population.[28]

Similarly, in Thailand many services are far better in Bangkok than in smaller cities and rural areas: Population per hospital bed in Bangkok was 3 to 12 times lower than that of cities with populations greater than 50,000 in the mid-1970s; Bangkok had 1 doctor per 1,900 people, compared to 1 doctor per 5,454 to 76,603 people in secondary cities. Nearly 85 percent of the population in the capital had access to electricity, but in all other cities (except Chiangmai) only 32 to 55 percent had electricity. Nearly 80 percent of Bangkok's population was served by piped water, compared to 28 to 61 percent in smaller cities.[29]

In many countries the large gap between the national capital and secondary urban centers in economic activities, job opportunities, and the quantity and quality of social services and facilities creates a vicious cycle that maintains secondary cities in a relatively weak economic position and makes the national capital even more attractive to professionals, skilled workers, and entrepreneurs. Analysts observed the adverse effects of the inferior quality of services, infrastructure, and facilities in secondary cities of Colombia:

> Why should doctors, or any professional, want to reside in an intermediate city when there are, for example, far superior educa-

tional, housing, sanitary, recreational and cultural facilities in the major cities? And without these professionals how can intermediate cities develop? This is not to suggest that all intermediate cities should have public facilities that rival those of the largest cities. However, the evidence available suggests that intermediate cities are so far worse off at this point in time and that their future development is threatened.[30]

Conclusions

The foregoing analysis implies that if governments in developing countries want to bring about a more balanced distribution of urban population and productive economic activities, as was done in South Korea, the economies of secondary cities must be made stronger through investments that support and foster industrial and commercial development. Such changes, however, must be based on a better understanding of the reasons that secondary cities have emerged and of their potential advantages, the dynamics of their growth, and the critical functions they can perform in the national economy. It is to these factors that the analysis turns in the next two chapters.

NOTES

1. Abukasan Atmodirono and James Osborn, *Services and Development in Five Indonesian Middle Cities* (Bandung: Institute of Technology, Center for Regional and Urban Studies, 1974).

2. James Osborn, *Area Development Policy and the Middle City Under the Indonesian Repelita Compared to the Malaysian Case: A Preliminary Analysis* (Santa Barbara, CA: Center for the Study of Democratic Institutions, 1974), pp. 39-40.

3. See Brian J. L. Berry and Frank E. Horton, eds., *Geographic Perspectives on Urban Systems* (Englewood Cliffs, NJ: Prentice-Hall, 1970), for a review and further references.

4. Harold Lubell, *Urban Development Policies and Programs,* Working Paper for Discussion, Bureau for Program and Policy Coordination (Washington, DC: USAID, 1979), p. 34.

5. See United Nations, Department of International Economic and Social Affairs, *Patterns of Urban and Rural Population Growth* (New York: United Nations, 1980), Population Studies 68, Table 48.

6. Ibid, p. 44.

7. Kingsley Davis, *World Urbanization 1950-1970*, vol. 1 (Berkeley: University of California, 1969), Tables B, C.

8. Robert W. Fox and Jerrold W. Huguet, *Population and Urban Trends in Central America and Panama* (Washington, DC: Inter-American Development Bank, 1977), p. 21.

9. Graeme Hugo, "Patterns of Population Movement in Intermediate Cities in Indonesia: An Overview and Some Issues and Examples" (Paper prepared for Workshop on Intermediate Cities in Southeast Asia, East West Center, Population Institute, Honolulu, 1980), p. 16.

10. See Visid Prachuamob and Penporn Tirasawat, *Internal Migration in Thailand, 1947-1972* (Bangkok: Chulalonghorn University, 1974), pp. 36-39; Ronald C.Y. Ng, "Recent Internal Population Movement in Thailand," *Annals of the Association of American Geographers* 59 (December 1969): 710-730; World Bank, *Urbanization Sector Working Paper* (Washington, DC: Author, 1972), p. 80; S. V. Sutharaman, *Jakarta: Urban Development and Employment* (Geneva: International Labor Office, 1976), p. 93; Government of India, Department of Statistics, *Statistical Abstract of India, 1974*, New Series 20 (New Delhi: Ministry of Planning, 1975), Table 3.

11. Davis, *World Urbanization*, Tables B, C.

12. Salah El-Shakhs, "Urbanization in Egypt: National Imperatives and New Directions," in *Development of Urban Systems in Africa*, ed. R. A. Obudho and S. El-Shakhs (New York: Praeger, 1979), pp. 116-136.

13. Richard Ulack, "The Impact of Industrialization upon the Population Characteristics of a Medium-Sized City in the Developing World," *Journal of Developing Areas* 9 (June 1975): 203-220.

14. Leighton W. Hazelhurst, "The Middle Range City in India," *Asian Survey* 8 (July 1968): 540.

15. Ibid.

16. A. A. Srivastaba, "Growth, Morphology and Ethnic Character of Ranchi-Dhurwa Urban Complex," in *Urbanization in Developing Countries*, ed. S. M. Alam and V. V. Pokshishevsky (Hyderbad: Osmania University Press, 1976), p. 539.

17. Fu-Chen Lo and Kamal Salih, "Growth Poles and Regional Policy in Open Dualistic Economies: Western Theory and Asian Reality," in *Growth Pole Strategy and Regional Development Policy*, ed. Fu-Chen Lo and Kamal Salih (London: Pergamon, 1978), pp. 191-234.

18. See Frederick Temple et al., *The Development of Regional Cities in Thailand* (Washington, DC: World Bank, 1980).

19. See Robert A. Hackenberg, *Fallout from the Poverty Explosion: Economic and Demographic Trends in Davao City, 1972-1974* (Davao City, Philippines: Davao Research and Planning Foundation, 1975), pp. 5-7; idem, "A

Retrospective View of the Poverty Explosion: Some Results of the 1979 Davao City Survey" (Paper prepared for Meeting on Intermediate Cities, East-West Center, Population Institute, Honolulu, 1980), Table 11.

20. Hackenberg, "A Retrospective View," p. 7.

21. Hackenberg, *Fallout from the Poverty Explosion*, p. 4.

22. See Akin L. Mabogunje, "Growth Poles and Growth Centres in the Regional Development of Nigeria," in *Regional Policies in Nigeria, India and Brazil*, ed. A. Kuklinski (The Hague: Mouton, 1978), pp. 3-96.

23. Cited in Harry W. Richardson, "An Urban Development Strategy in Kenya," *Journal of Developing Areas* 15 (October 1980): 97-118.

24. B. Jenssen, K. R. Kunzmann, and S. Saad El-Din, "Taming the Growth of Cairo: Towards a Deconcentration of the Metropolitan Region of Cairo," *Third World Planning Review* 3 (Spring 1981).

25. Sam P. S. Ho, *Small Scale Enterprises in Korea and Taiwan*, World Bank Staff Working Paper 384 (Washington, DC: World Bank, 1980).

26. Atmodirono and Osborn, *Services and Development*, p. 108.

27. U.S. Agency for International Development, *Colombia: Urban-Regional Sector Analysis* (Bogotá: Author, 1972), p. 46.

28. Alan Gilbert, "The Spatial Allocation of Education and Health Facilities in a Less Developed Nation," in *Proceedings of the Commission on Regional Aspects of Development of the International Geographical Union*, vol. 2, ed. F. Helleiner and W. Stohr (Ontario: IGU, 1974), pp. 307-344.

29. See Temple, *Development of Regional Cities*, pp. 16-17.

30. U.S. Agency for International Development, *Colombia: Urban-Regional Sector Analysis*, p. 46.

CHAPTER 3

DYNAMICS OF GROWTH IN
MIDDLE-SIZED CITIES

Despite their relative weaknesses in the economies of developing countries, secondary cities seem to perform important economic and social functions that can contribute to national development. Attempts to encourage the growth and manage the development of secondary cities must be based on a better understanding of these functions and of the reasons that these urban places emerged as larger and more diversified economies. The dynamics of growth in secondary cities has been given relatively little attention in scholarly and professional literature. Thus this chapter reviews the factors that have influenced the founding and early growth of contemporary secondary cities through the late 1950s, and explores the dynamics of their development.

Most of the secondary cities in developing countries reached 100,000 population only recently. In 1950, less than 5 percent of all urban places in the world had populations of that size. The number of secondary cities in developing nations that year was less than half the current number.[1] But urbanization is a recent historical phenomenon for all countries. Before 1900, less than 13 percent of the world's population lived in cities. In 1800, for instance, only about 5 percent of the population could be found in settlements with more than 5000 residents. As late as 200 years ago, only 1 city—Peking—had reached a million in population, and only 6 others had more than a half million people. It was not until

---

Portions of this chapter appeared in slightly different form in *The Geographical Review*, Vol. 73 (1983), and are reprinted with the permission of the American Geographical Society.

the mid-1800s that any city in Europe reached 1 million in population, and by 1900 only 16 cities had grown to that size. Until the 1800s most of the world's largest metropolises would be considered small or secondary cities by contemporary standards.[2]

The present patterns and scale of urbanization in the West were born of the industrial revolution. Urbanization did not begin in Europe until the end of the nineteenth century, and the highly concentrated pattern of urban growth now seen in many developing nations did not appear until the early twentieth century, mostly in countries colonized by European nations. Differences in the distribution of urban population in Western industrial countries, where it was more balanced among large and intermediate cities, and the developing countries—where most urban dwellers were concentrated in one or a few large metropolitan centers—are usually attributed to colonial economic policies or to attempts by postindependence regimes to promote rapid industrialization. Industrialization and urbanization seem to have been concomitant phenomena in Europe and North America, each reinforcing the growth and spread of the other. But in the colonies, only one or two cities were allowed to grow to large size, usually to serve as entrepots for exporting raw materials and importing manufactured goods. Smaller towns and cities were encouraged to grow only if they were convenient administrative centers through which colonial governments could extend political control over the hinterlands, or if they could serve as transfer points in transportation systems designed to exploit raw materials and agricultural resources in interior regions. Little attempt was made in most developing countries to create a system of secondary cities that would generate demand for domestically produced goods or make urban services and facilities available to a large majority of the population. The networks of secondary cities that emerged with, and helped to promote, widespread industrialization in Western nations failed to appear in most developing countries, either because the spatial implications of national investment policies were ignored or because the policies were deliberately designed to contain industrial and commercial activities in one or a few major cities.[3] Thus the largest metropolises became the only feasible locations for the heavy concentration of investment in infrastructure, technology, and plant needed to promote rapid industrialization when that became the primary goal of development policies in the 1940s and 1950s. These metropolitan centers continued to grow to enormous size in many Latin American and Asian countries during postindependence periods. Secondary city growth in those regions, and in much of the Middle East, however, is largely a post-World War II phenomenon.

Although a majority of contemporary secondary cities reached 100,000 population only after World War II, some are ancient settlements. The forces that stimulated their growth prior to the 1960s were similar for both ancient and modern settlements. Among the most important were: their favorable physical location and endowment of natural resources; their selection as political or administrative centers; the concentration in them of colonial or foreign investment; conditions favorable to making them commercial and service centers for their regions; the influence of transportation routes and technology; and the impact of government investment in infrastructure and facilities. As will be seen later, nearly all contemporary secondary cities grew as service centers of one kind or another; indeed, their rationale for existence in most cases was that they provided easy access to commercial or personal services. Before 1960, large-scale manufacturing had influenced the growth of relatively few secondary cities in developing nations. To the extent that they had significant manufacturing sectors, they were composed primarily of small and medium-sized enterprises with low levels of capitalization, usually employing less than 10 workers.

Many factors that influenced the growth of secondary cities before 1960 still shape their development and continue to influence the growth of smaller towns; these factors require more detailed description.

Favorable Physical Location and Natural Resources

In his study of urbanization in developing countries, Breese points out that "it is noticeable in urbanization everywhere that the factors of site and situation have considerable impact upon the nature of the urban area."[4] Physical location (site) and the relationships among sites (situation) have controlled the growth of cities throughout the developing world. Case histories of secondary cities in developing nations repeatedly identify favorable physical characteristics and natural resource endowments as distinguishing factors in their development. Physical and natural endowments were instrumental in the founding of these settlements, in making them attractive as commercial and service centers, and in diversifying their economies later in their history.

Location near the sea, for example, has stimulated city growth in nearly every geographical region of the world. Most of the largest cities in developing nations are seaports. Locations at the junctions of navigable rivers and along the coasts of natural bays have spawned secondary cities as well. For instance, coastal locations were vital to the founding and expansion of Matamoros, Tampico, and Veracruz in Mexico, Barranquilla and Cartagena in Colombia, and Belem, Fortaleza, Recife, Salvador, and Santos in Brazil. The Korean cities of Incheon, Pohang, Kunsan, Ulsan, and Pusan grew in large part because of their access to the sea. Observers note that in Malaysia, the "domination of the colonial ports over smaller, coastal settlements was an important feature in the development of the urban hierarchy."[5] The coastal cities not only developed more rapidly, but their size and economic diversification were closely related to the volume of port activities. Surabaya and Padang in Indonesia, Georgetown in Malaysia, and Madras and Calicut in India grew into significant intermediate cities as a result of their location near the sea. In the Middle East, Tripoli emerged as an important Lebanese secondary city because of its port. African secondary cities grew first within a band of about 150 miles along the coasts, and coastal location shaped the direction and pace of growth in port cities. DeBlij points out in his study of Mombasa, Kenya, that "for centuries development here was based upon the use of Mombasa Harbor," with its 300-yard-wide entrance channel.[6] Urban development occurred first on the periphery of the port and later expanded outward. The influence of the seaport was seen clearly in the late 1800s, when economic activities were transferred from Mombasa Harbor to a smaller port, Kilindini, on the western side of the island, and the direction of Mombasa's growth shifted dramatically toward the west. Commerce and trade began to grow on the outskirts of the city, quickly producing new sections that were clearly distinguishable from the "Old Town."[7]

However, most secondary cities in developing countries are inland, and favorable physical and environmental resources were important to their growth. Studies of Tlemcen, Algeria, conclude that its location on a narrow plateau in the slopes of a mountain overlooking a broader plain provided it "a particularly favorable position at the point of contact between two contrasting but complementary zones," and that "the continuity of urban life from Roman times to the present day owes much to the varied natural resources of the surrounding area."[8] Tlemcen's hinterland, a broad plain lined with springs and streams and forested with oak and juniper, had soils conducive to commercial agriculture and the raising of livestock.

Ranchi, India, was no more than a small tribal village in the early 1800s, and its growth to over 100,000 population in 1950 and to more than 400,000 in 1980 is attributed largely to its location on a plateau in a hilly region of Bihar on the River Harmu. Its central location made it an ideal administrative center in the early 1900s, and it easily accommodated the commercial and industrial activities that were attracted there in later years.[9] Similarly, Chiangmai, Thailand, has enjoyed the advantages of its location on a flat plateau near the source of the Mae Ping River since the early 1300s. The city's moderate climate and fertile soil allowed farming to flourish and made it a strategic defensive position in the continuing battles among the Burmese, Laos, and Thais from the 1500s to the 1700s, and the same features made it a convenient site for regional administrative and trade functions from the 1800s to the present.[10]

Taichung, one of Taiwan's five largest secondary cities with a current population of more than half a million, was, like Ranchi, a small farming village and periodic market in the late 1800s, having less than 1500 residents. Its location on a broad alluvial plain between two of Taiwan's largest rivers combined with its hot, wet summers and cool, dry winters to provide ideal farming conditions. Its central location in a rich agricultural region made it a natural center of trade and, later, an important defensive and administrative node. Pannell observes that "Taichung's site and locational characteristics were seen as vital to the city's historical development and resulted in rapid rates of population growth and areal expansion."[11]

Mineral and metal resources were no less important in the creation of many secondary cities in Africa and Latin America. Copperbelt towns in Zambia grew around mining and refining operations, and the mining industries were instrumental in linking interior settlements by rail or road to larger cities, generating opportunities for commerce and trade in the mining towns. One of these mining centers, Kitwe, grew in population during the 1960s to rival Zambia's capital, Lusaka.[12] Similarly, Medellín, Colombia's second largest metropolis, grew as a service center for gold-mining activities in the Antioquia region during the sixteenth and seventeenth centuries. The natural resources and favorable climate were also conducive to coffee growing, and during the 1800s the city's rapid expansion was stimulated by coffee trade. Coffee cultivation stimulated internal consumer markets, generated capital for investment in manufacturing, and provided the base for the industrialization that took place later in Medellín and its metropolitan area.[13]

Defensive Positions and Military Bases

Physical characteristics also gave some secondary cities advantages as defensive positions and military bases. Its location near the sea, for instance, made Tripoli, Lebanon, an important citadel during the thirteenth- and fourteenth-century crusades and a base of military operations during the time of the Ottoman Empire.[14] Chiangmai achieved its status as an important secondary city in Thailand as early as the thirteenth century because its location provided a strategic defensive position that made it a prize of conquest in succeeding battles among Thailand and its neighbors. Breese notes that in India military camps had profound effects on rates and patterns of urban growth during the twentieth century. "Where they are longitudinal to main travel arteries to the central city, they may not substantially alter growth direction but merely withhold land from urban expansion," he observes. "Where they lie athwart expansion, they may either deflect or halt it." To the extent that the land for military camps was convertible to other uses, it shaped the direction of urban expansion in smaller and intermediate Indian cities.[15]

The selection of African towns and cities as defensive positions or military bases also had a profound influence on their growth. Most of Morocco's secondary cities were founded as fortifications. Their crenellated walls and towers are still visible and now divide ancient parts of these cities from their modern accretions. For centuries, the fortifications shaped the physical growth of these cities and dictated the location of social and economic activities. In his analysis of urbanization in Nigeria, Ajaegbu observed that most of the cities and larger towns that grew to significant size before 1900 did so because of the protection they offered from tribal warfare and the base they provided for military conquest. Yorubas, Binis, and Fulanis sheltered people from the countryside in their cities, protected them against the forages of colonial armies, and used the cities as staging areas for attacks on other tribes. "The original impetus to the growth of the city of Zaria happened by means of the protective powers of her warrior rulers over the vast hinterland of the then Hausa kingdom of Zazzau," Ajaegbu recounts, "and by means of the war conquests of and tributes paid to the city and her rulers from as afar afield as Kano, Katsina, Bauchi and the Nupe country to the south up to the Niger River."[16]

The residents of Ibadan—a Yoruba military camp in the early 1800s—were later able to turn that city into an important trading center and thus ensured its continued growth when it no longer served a defensive purpose.[17]

Many cities that reached significant size because they performed other administrative or economic functions also received temporary stimulation from military activities. Their selection as temporary military encampments or headquarters during wars often generated rapid, and permanent, population growth. Huancayo, a small secondary city in the Mantaro Valley of Peru, provided a convenient location for a Peruvian military camp during the wars with Chile in the late 1800s; this increased both its population and its functions as a commercial and service center.[18] The selection of Ranchi, India, as the headquarters of the Eastern Command during World War II also increased its permanent population by nearly 70 percent, added to its already important administrative functions, and helped diversify its economy, paving the way for the location of modern manufacturing activities there during the 1950s and 1960s.[19] Similarly, the Palestinian War in the Middle East in the late 1940s brought Tripoli, Lebanon, unexpected economic growth when it forced the headquarters of the Iraq Petroleum Company out of Haifa. The company relocated in the relatively more secure city of Tripoli, and brought with it 2000 new families, who created a building boom and generated demand for furniture, appliances, and a wide range of other consumer goods and services. Gulick wrote that "this chain of events was the greatest single impetus to physical growth [in Tripoli] since World War II."[20]

Selection as Administrative and Political Centers

A set of factors similar to those that have stimulated the growth of national capitals also seems to have promoted development of towns and cities selected as regional, provincial, or state capitals or as local administrative or political centers. Those places designated as administrative centers have benefited greatly from investments in infrastructure and from the commerce and services attracted to places with political status. Abu-Lughod observed that during the 1950s and 1960s nearly all of the cities between 100,000 and 500,000 population in the Middle East were provincial capitals. Damanhur, Asyut, Mansura, Giza, Zaqaziq, Tanta, and other secondary cities in Egypt were province capitals, as were Tripoli, Lebanon; Zarqua, Jordan; Homs, Harna and Lattaqiya in Syria; and Kirkuk, Mosul, Basra, and Hilla in Iraq. Nearly all other provincial capitals ranged from 40,000 to 100,000 in population, and there was a large gap in population size between these provincial cities and the average town or village in the next city-size category.[21] Similarly, Weiker has found that the most economically important and fastest growing secondary cities in Turkey have been provincial capitals. He notes that "the potential for Turkish provincial cities making important contributions to national economic, social and political development is great." He observes that among their assets were the "attractiveness of the provincial cities for migrants from rural areas; increasingly skilled, capable and enthusiastic city leadership (although much remains to be done in this area); a striking degree of local spirit, and frequently vigorous and creative responses to challenges."[22]

The investments that come with the designation of a city as a regional or provincial administrative center seem to have been important in their growth in most regions of the developing world. In his study of Parana in Argentina, Reina insists that "the strongest impulse to development came when Parana became a capital of the province in 1883," for only then was more serious attention given to physical improvements and to providing basic services. Investment in infrastructure by both the local and central governments increased markedly.[23] Ghana's second largest city, Kumasi, founded in the late 1600s as the capital of the Ashanti Confederacy, was not only an important political center but also emerged as a crossroads of commerce and trade. Its role as an administrative capital was reinforced by British colonialists, who designated it as headquarters of the Northern Territories. As a regional capital it was connected by rail and road to Accra, Ghana's capital, and as a result of its administrative status Kumasi's population grew from less than 25,000 in 1921 to more than 380,000 in 1970.[24]

In Asia, traditional cities were always administrative centers through which warlords and emperors maintained political control over, and extracted revenues from, the hinterlands. Most were neither founded nor functioned as commercial or manufacturing centers. As Murphy has pointed out, "the traditional Asian city was predominantly a political and cultural phenomenon rather than an economic one."[25] Only after the penetration of European traders did urbanization in Western style begin in Asia, first at port cities and then at inland towns connected to them. Even informal political functions stimulated growth in secondary cities. The selection of Chiangmai as the royal family's palace, for instance, assured preference for the northern Thai city in the allocation of investments for modern services and facilities, infrastructure, and road and airport facilities. The prestige and status accorded to Chiangmai as a place in which the king entertained important government officials and international visitors made it a national, and later an international, tourist attraction.[26]

Political ideology seems to have little effect on the preference governments give to regional and provincial capitals for development investments. Even in China during Mao's regime—one noted for its deliberate attempts to make development more equitable—the provincial capitals fared better in economic growth and accumulation of services and facilities than noncapital cities. Provincial capitals consolidated, and often increased, their primacy in provincial settlement hierarchies during this period. Reviewing the changes in urban development in China between 1953 and 1972, Sen-Dou Chang notes that provincial capitals throughout the country and especially in the rural northeast and southwest grew rapidly. Even taking boundary changes—which explain some of the growth in urban population in China—into consideration, and "despite the extreme variations in physical conditions, resource endowments, population densities, transport facilities and agricultural productivity in the various provinces and autonomous regions," Chang observes, "the growth of the capital cities throughout the country seems to be uniformly fast."[27] As a result of the higher levels of investment in provincial capitals, these secondary cities gained advantages from better infrastructure and services, and from their nodality in the transport system, to make them

more efficient locations for heavy and light industrial investment in subsequent years.

Colonization and Foreign Investment

A related, and sometimes equally important, factor in the
founding or early growth of secondary cities has been the
activities of foreign investors and traders and the policies of
colonial regimes. Although unbalanced development of urban
systems in developing countries and the overconcentration of
industry and modern infrastructure in national capitals are
often attributed to colonial economic policies, these policies
often stimulated the growth of secondary cities as well. In
some cases secondary cities were encouraged to grow as
colonial administrative posts or as transfer and processing
centers for exploitation of mineral and agricultural resources
in the interior.

In some Southeast Asian countries, for example, there was
no indigenous urbanization before European colonization. In
Malaysia, the major cities and towns are less than 200 years
old and did not begin to grow rapidly until the early 1900s.
The colonial economic system had a strong and pervasive
influence on the development of towns and villages in Malaysia. Cities emerged in those areas that could grow crops for
export or that had endowments of exportable natural
resources; agricultural centers and mining towns were connected by rail and road to port cities and grew as processing
and transfer centers. Those colonial administrative posts that
could attract commerce and trade diversified their economies; those that could not declined in the period after
independence.[28]

Taichung, Taiwan, grew rapidly into a secondary city only
after Japanese occupation leaders chose it as a civil affairs
administrative center in the late 1800s. Because of its central
location, it became an important part of the Japanese admin-
istrative-defensive network in Taiwan in the early 1900s.

Colonial administrators invested heavily in roads, bridges, water, and drainage systems. They linked Taichung to other cities and towns. The grid street pattern they laid out ordered the physical growth of the city for years afterward. By 1920, the city had twice the population it had at the beginning of the occupation.[29]

Colonial governments—Spanish, Japanese, and American— also profoundly influenced the economies of secondary cities in the Philippines. The foundation of the present economy of Davao City, for instance, was built during the Japanese occupation. Until then, Davao had a subsistence agricultural economy with little manufacturing or commerce. The Japanese created a plantation agricultural system, built roads, refurbished port facilities, and fostered the export of abaca, copra, and timber, all of which stimulated the city's subsequent population growth and economic diversification.[30]

In Latin America, Spanish and Portuguese colonists played a strong role in founding and promoting the growth of contemporary secondary cities. Morse notes the strong control that Portuguese rulers had over the development of towns in Brazil, observing that "Brazilian towns were, like the Spanish, products of the metropolitan [colonial] will. . . . Nothing spontaneous or natural attends their birth. For some, even the site is preselected from Lisbon."[31] But in most of Latin America, colonial regimes were more interested in establishing towns and cities for specific economic or administrative purposes than in stimulating the growth of a network of secondary urban centers. Morse argues that, as a result,

urban networks developed feebly. Geographic barriers to regional transportation were often formidable, while the crown's mercantilist policies did little to encourage centers of complementary economic production. New World cities tended to be related individually to the overseas metropolis and isolated from one another.[32]

Foreign influences also played an important role in the early growth of secondary cities in Africa and the Middle East. Tripoli grew in importance with successive waves of foreign traders who came in the seventeenth and eighteenth centuries seeking silks, citrus fruit, olives, vegetable oils, and tobacco.[33] Lawless and Blake point out that the French occupation of Algeria and the activities of colonists in and around Tlemcen created "profound social, economic and geopolitical changes—the arrival of a privileged European minority, the dramatic increase in the size of the urban population, the appearance of new functions, the disappearance of some traditional functions, and the strengthening or transformation of others."[34] The French built fortifications at Tlemcen, and for much of the early 1800s it served as a garrison; the walls and barracks shaped the boundaries and directions of city growth.

The development of Malindi, Kenya, depended almost entirely on succeeding groups of foreign settlers and traders. Arabs established plantation agriculture in the Malindi area in the late 1800s using slave labor and were followed by Asian merchants, who made the city a regional wholesale and retail trade center. In the 1930s, Europeans settled there in large numbers to open hotels and shops catering to tourists, making tourism the major productive sector in Malindi's economy until after World War II.[35] A similar succession of Arab, Portuguese, Asian, British, and European colonists and traders also shaped the growth of Kenya's secondary metropolis, Mombasa.[36] As will be seen later, however, the foreign influences on the growth of secondary cities were not always beneficial to developing nations: Some cities became parasitic rather than developmental.

Influence of Transport Technology

From the middle of the nineteenth century onward, the introduction of new transport technology played a critical role in the founding, growth, and decline of secondary cities in the developing world. The expansion of these cities in the Middle East and Northern Africa was inextricably tied to the extension of road and rail networks. Tripoli, Lebanon, experienced a new surge of population growth and attracted new economic activities in the early 1900s when roads were opened to Beirut and when the railroad connected Tripoli to Aleppo in 1911. For many years to follow the physical expansion of the city followed the main transport routes.[37]

Transportation networks played an equally crucial role in the growth, spatial distribution, and functional development of Nigerian secondary cities. The roads and railways built at the turn of the century to exploit the resources of the interior often bypassed traditional trade centers such as Ile-Ife, Ilesha, Benin City, Sokoto, Katsina, and Yauri, and turned what were previously small villages into important nodal points in the colonial spatial and economic system. Kaduna, Jos, Enugu, and Port Harcourt, all located along the new rail lines, grew in population and became more economically diversified. Only the traditional centers that were integrated into the major rail and road system—Ibadan, Oshogho, Zaria, and Kano—were transformed into viable economies. These secondary cities took on dual structures, with attributes of both traditional trading centers and more modern commercial and service centers.[38] The completion of the railway to Ilorin in 1908, for example, increased the flow of manufactured goods to the city from Lagos and of agricultural products from the north and from the city's hinterlands. As in many other secondary cities in developing nations, commercial, trade, and administrative functions expanded in the area around the railroad stations. Akorede observes that "the railways and roads that accompanied the colonial administration were followed by developments in business and commerce."[39] Land values near Ilorin's railway station rose rapidly and the direction of the rails ordered the direction of the city's physical growth. New towns eventually sprang up at breaks in transport networks.[40] Gugler and Flanagan point out in their analysis of urbanization that "for West African towns, fortune rode the trains. Those that received terminals grew, but those that did not stagnated or declined, as did many river ports."[41]

Case histories of secondary cities in Latin America reveal similar influences of transportation on growth and decline. Different modes of transportation had different influences on the growth of secondary cities at different periods in their development. Extension of the railroad to Huancayo, Peru, in 1908 strengthened that city's role as a communications and trade center. Expansion of the highway network to include Huancayo in the 1940s and early 1950s contributed to a resurgence and diversification of the city's commercial activities and was instrumental in bringing textile mills that, at least temporarily, employed large numbers of workers who came from the countryside to supplement their farm incomes.[42]

The increased access brought by the highway systems in Latin America was especially important to smaller, interior cities. The growth and development of Oaxaca, Mexico, is reported to have stemmed directly from its access to the Pan-American Highway, which was extended to Oaxaca in 1940. The highway not only linked what was then a relatively small urban center to the Mexico City metropolitan area, but also strengthened Oaxaca's position as a regional market. Since the highway's construction, the city's population has grown by about 4 percent a year, and now it has more than 150,000 residents.[43]

Transport was no less important in stimulating the growth of secondary cities in Asia. Hazelhurst has observed that middle-sized cities in India are most commonly found along main road and railroad routes, especially in the area north of Delhi and in the west toward Punjab. The commercial and trade activities of these cities have been densely concentrated around rail and bus stations.[44] Roads and railroads also played a role in the growth of Chiangmai, Thailand, during the early part of this century. A railroad link constructed between Chiangmai and Bangkok and a highway completed in the mid-1930s were crucial in maintaining and strengthening the city's functions as an administrative center for the northern region and in allowing it to emerge as a commercial, trade, and tourist center later. The construction of intercity roads and highways was essential in forging linkages between Chiangmai and other northern towns and cities. Construction of seven radiating roads out of the city's center, and of the Chiangmai-Lampang highway, increased the volume of trade and the number of tourists to Chiangmai, and allowed it to become a vital service center for the northern region in the 1960s.[45]

The construction of roads from other Philippine cities to Davao in the 1940s also provided the access upon which much of that city's economic development during the next two decades depended. At the end of the Japanese occupation, a road was opened between Davao and Cotobato and later one was extended from Davao to smaller towns in Agusan Province that not only fostered intercity trade but opened interior crop and timber land to use by Davao entrepreneurs and farmers. Access to markets gave impetus to agricultural development wherever timberland was cleared. Davao City's population more than doubled, from 40,000 in 1948 to more than 80,000 in 1960, and the population of the province in which the city is located grew by more than 145 percent during the same period.[46]

Growth of Commerce, Trade, and Services

Without doubt the most important factor in promoting the growth of nearly all contemporary secondary cities was their capacity to perform commercial, service, and trade functions. Many began to grow initially because of their central locations in the midst of rich agricultural lands, or because they were selected as administrative centers or defensive positions, or because they were linked to other cities by road, rail, or water, but their continued growth and diversification depended on their ability to foster commerce and trade. Many of the secondary cities in Latin America grew from agricultural trade and acted initially as regional markets and transfer points for agricultural goods going from their rural hinterlands to larger metropolitan centers. Huancayo, Peru, began to grow in the late 1800s because of rising demand in Lima and other coastal cities for agricultural commodities grown in the Mantaro Valley. Land in the coastal cities had been steadily converted to plantation farming for export crops, leaving domestic urban markets dependent on agricultural goods grown in interior regions. The commercialization of agriculture in the Mantaro Valley not only made Huancayo an important agricultural market, but also stimulated diversification and established the city as a commercial and trade center for its region. Its growth in the late 1800s was directly related to the expansion of commerce and services. "By the end of the century," Roberts recounts, "Huancayo was mainly a commercial town."[47] An inventory of commercial establishments in Huancayo taken in 1880 listed Chinese restaurants, a hotel, lodging houses, ice-making, ham, and liquor factories, 52 shops in the permanent market, general stores, leather goods stores, shoemakers, and barbers.[48] Flour mills had been established, commerce in eggs was brisk, and wholesale traders from other cities passed through Huancayo frequently. By the 1920s the city had acquired banking and financial services, and was a regional judicial and communications center through which labor was recruited for large haciendas and nearby mines. The city's population expanded again in the early part of the twentieth century, when entrepreneurs established large-scale contracting services to supply mining companies with goods and equipment. Trucking and transport services to ship out agricultural products and ore quickly established themselves in Huancayo. Repair shops to maintain vehicles and equipment sprang up just as quickly.[49]

The preindustrial commercial and service centers of Africa could support substantial numbers of residents, and this undoubtedly accounted for their rapid population growth. Commerce in Ibadan, Nigeria, at the turn of the century involved large numbers of small-scale establishments that yielded small but widely distributed increments of new income. Muench points out in his study of Ibadan that the wide distribution of income generated through small-scale commercial and manufacturing activities allowed the city to grow because it created demand primarily for small, inexpensive, locally produced consumer goods. Even imported foods and materials were processed by local enterprises to keep costs down—meat shipped from other regions was processed by city butchers; furniture was made by local craftsmen from wood brought in from outside; local tailors made clothes from imported textiles; and metals brought into the city from other areas were fashioned into jewelry by local smiths. Large numbers of small-scale wholesalers and retailers traded a variety of agricultural commodities, livestock, and crafted goods. The wholesale and retail activities, in turn, supported an extensive and multitiered system of consigners, brokers, distributors, and other "middlemen" who broke up consignments several times to allow even petty traders and hawkers to participate in commerce. According to Muench, "This increase in the incomes of primary producers in the city has tended to bring a larger increase in population than has been the case in many other African cities. This multiplier effect has been an important factor in the growth of the city to its present large scale."[50]

Secondary cities in Algeria grew as agricultural markets and diversified into commercial and service centers as farmers turned more of their fields to the production of cash crops. During the French colonial period, when both domestic and foreign demand for local production was high, larger market towns became regional centers for banking, transport, storage, and wholesale and retail trade, and, depending on the number of foreign residents, for luxury goods shops and import trade. After a period of settlement by colonists, agriculture was transformed almost entirely to cash-crop production, transforming the economies of market towns into trade and service centers. Lawless and Blake recount the process in Tlemcen in the late 1800s: As agriculture became more commercialized,

> the town acquired new service and marketing functions. It became the most important centre for transport and storage, banking and wholesale and retail trade in its region. The town ceased to be the exclusive market for the agricultural products of its region and became also the intermediary between the rural areas around and the capitalist market. It became the instrument by which agricultural products were collected for export to a greatly enlarged market at both the national and international scale. These central place functions resulted in the establishment of an impressive range of public buildings in Tlemcen including, for example, banks, post offices, and hotels, besides warehousing and the railroad station. Cultural and social facilities and schools were also built for the benefit of the French community, and Western type shops, offices and cafes sprang up near the centre of town. The extent to which the town was frequented as a regional centre depended primarily on geographical factors—accessibility and distance.[51]

As in Ibadan, manufacturing in Tlemcen has nearly always been small in scale. Shops producing carpets, carved hunting guns, copperwork, iron and clay pots, wool and leather goods, and jewelry flourished in Tlemcen in the late 1800s and early 1900s.

The social services provided in secondary cities were no less important than commerce and trade in stimulating their growth. The concentration of primary and secondary schools in Tlemcen in the early 1900s made the city an educational center for its region, and complemented its commercial activities. Observers point out that "in 1890, the Ecole de Pais des Indigenes was founded and continued to have an important influence on carpet making and design in Tlemcen for much of the French period."[52] Some cities were especially influenced by the social services established by foreign missionaries. Wherever they went, European and North American religious groups set up schools as well as churches, hospitals, orphanages, and social services that were used both to proselytize and to introduce new ideas and values that were to change the economies and societies of the towns in which they settled. Gulick notes that Christian missionary schools in Tripoli had an important impact on the Arab city's social structure in the late 1800s.[53] The schools, hospitals, and colleges built by missionaries in Ranchi, India, in the late 1800s and early 1900s also contributed greatly to the city's emerging role as an administrative center. Srivastaba observed that the mission schools affected the economy of the city in a number of ways. They were noted for "high standards of discipline, education, service and devotion to work," he argues, and thus became "centers of diffusion of modern and western civilization into tribal life and culture." Whatever the desirability of such influences on the poor in Ranchi, these services "increased their employment opportunities, social status and functional capabilities," according to Srivastaba, and contributed to making Ranchi an urban center to which both government expenditures and private investment would later flow.[54]

The concentration of government investment in services, facilities, and physical infrastructure at certain periods seems to have been a major influence on the growth of secondary cities. They were important in consolidating the economic advantages of these cities, in enhancing the value of previous investments, and in encouraging new private investment. Often they created conditions that allowed diversification into entirely new economic activities.

Dotson and Teune's factor analysis of Philippine cities shows an extremely strong correlation between population size and the concentration of industrial, service, and commercial establishments in those cities. During the period from 1939 to 1960, larger cities in the Philippines, compared to smaller, less urbanized places, had more manufacturing and service establishments per capita and higher levels of educational attainment, at least in part because governments in larger municipalities generally raised more revenues and spent more for public purposes.[55] These comparative advantages in social services and facilities have often been crucial in attracting population and investment that allowed secondary cities to continue to grow.

Dynamics of Secondary City Growth: The Importance of Synergism

Although factors that were crucial in the founding and early growth of secondary urban centers in developing nations can be identified and described, it is clear from historical studies of their development that these initial influences can only partially explain their continued expansion and diversification. Some of the forces that generated growth during their early history were random. Although some cities were planned and their physical growth was carefully designed, the large majority of secondary cities grew from spontaneous actions by individuals reacting to favorable conditions. As some of these initial forces weakened, those cities that could not transform and adjust declined or disappeared. Perhaps the most dramatic example is that of Potosi, now a relatively poor Bolivian town of less than 100,000 residents, that in the mid-1600s was the largest city in the Western Hemisphere and one of the richest mining centers in South America. Its population of about 160,000 was larger than Mexico City's at the time, and its wealth was known throughout Latin America and Europe. Chapman reports that in 1556, a scant eleven years after its discovery, the city spent eight million dollars (pesos) to celebrate the accession of Philip II to the [Spanish] throne. In 1580, it was estimated that private fortunes of the dominant class ranged from about three hundred thousand dollars to six million. . . . At one time in this city there were fourteen dancing schools, thirty-six gambling houses and one theatre, to which the price of admission ranged from forty to fifty dollars.[56]

Potosi had one of the most productive mints in Latin America, supplying coins and medallions to Bolivia, Spain, and many of the Spanish colonies. Its churches, art collections, and public buildings were magnificent. But the economy of the city depended almost entirely on mining and on the slave labor of indigenous Indians; when the mines were no longer productive, the city's population declined almost as quickly as it grew. By 1825, it had less than 10,000 residents. Now Potosi is a minor administrative and commercial center, showing few signs of its former wealth except in its old and deteriorating buildings. Its residents, and the Indians living in its rural hinterlands, are among the poorest in South America.

Factors that were important to the growth of some contemporary secondary cities were detrimental to that of others. In Latin America, for instance, many of the secondary cities in the interior that grew as service and resource extraction centers in the eighteenth and nineteenth centuries, like Potosi in an earlier period, either stagnated or declined after modern transportation facilities put them in competition with coastal ports and national capitals in the late nineteenth and early twentieth centuries.[57]

Synergism was crucial in the dynamics of growth and development of secondary cities. Nearly all of the studies reviewed here show that a combination of influences stimulated the growth of these cities and that none of the factors that were important in their development could, alone, have generated growth and diversification. Studies of cities in

Southeast Asia and Latin America show that unless those that were founded as mining or port towns could diversify, be linked physically and through trade and social interaction with smaller towns and larger cities, begin to perform central functions, consolidate their economic and social gains, create comparative advantages, and respond quickly to external and internal social, political, and economic changes, they did not continue to grow. Observers note that in Malaysia

> the towns which evolved from mining settlements did not perform central functions. Rather they performed special functions with peculiar locational demands. . . . It was not until the settlements acquired administrative functions that there was any element of stability, let alone permanency. It was not until the introduction of such facilities as communications that the settlements had relations with their neighbors. There was little impetus for development of trade other than what was required for the immediate needs of the mining community.[58]

A complex combination of forces differentiated those towns that achieved a minimum level of urbanization and then stagnated or declined from those that grew into secondary cities. Among the factors that seem to have been most important were:

(1) initial growth factors
(2) reinforcing influences
(3) consolidating forces
(4) linkage effects—internal and external
(5) creation of comparative advantages
(6) agglomeration and multiplier effects
(7) new diversifying influences
(8) "second-generation" reinforcing factors
(9) new consolidating forces
(10) broader linkages and networks of exchange
(11) greater agglomeration and multiplier effects
(12) stabilization of some initial growth factors

(13) decline of some initial advantages
(14) size-ratchet advantages—protection against drastic decline
(15) new diversifying influences

Initial growth factors such as favorable location and resource endowments, selection as an administrative or political center, advantages as a defensive position or military base, the concentration of foreign trade or colonial investment, favorable location for regional trade and commerce, government investment in infrastructure and social services, or location along an important transport route often mutually reinforced each other during early stages of development, and allowed those settlements to consolidate their specialized activities. Linkages among activities within these cities and with other towns allowed them to attain comparative advantages in the performance of some functions, most often trade, administration, or services, which attracted migrants from rural areas and created more opportunities for investment in productive and service activities. New diversifying influences were needed later, however, to allow these cities to absorb increasing numbers of population. Changes in transportation routes, the application of new technology, the influx of foreign capital or the investment of locally produced surplus capital in complementary economic activities, the accumulation of new administrative functions, or surges of government investment in infrastructure and services created new conditions that allowed more diversified economic and social activities to survive in these cities. Larger population size and greater economies of scale and proximity allowed them to perform new functions. These new diversifying influences again required reinforcements through social, economic, physical, and political transformation, and the consolidation of these gains through institutional arrangements. As their economies became more diversified, these cities could broaden their linkages and interactions with other cities and towns and expand their networks of exchange and trade. Greater agglomeration produced new multiplier effects.

Some of the forces that initially influenced development became weak or less important, while others stabilized and continued to promote growth and diversification. As cities grew to from 250,000 to 500,000 in population, they achieved what Thompson has called "size-ratchet" advantages.[59] Large population size and economic diversification gave them some immunity to rapid or drastic decline, allowing them time to react to changes in economic and social conditions and to offset the effects of decline in any one sector. The continued growth and diversification of these cities depended on the repetition of this cycle of innovation, reinforcement, consolidation of gains, expansion of linkages, creation of new comparative advantages, higher levels of agglomeration, greater economies of scale, and reconsolidation of gains at new levels of development.

Merely growing to 100,000 or more in population, however, has not ensured that secondary cities would become catalysts for development of their regions. The spread of benefits to their hinterlands has often been constrained, and, indeed, some cities grew in such a way that they exploited their rural peripheries and drained resources from them to feed their own growth. Some secondary cities simply grew as enclaves, in which higher standards of living were achieved by a small group of residents and the vast majority remained poor. These cities were "exploitational." Other, "developmental," cities grew in such a way that linkages of mutually beneficial exchange were forged with smaller towns and rural areas in their regions, and these secondary cities not only provided higher levels of income and greater access to services and facilities to their own residents, but also to people living in the region in which the city was located. The factors that seem to differentiate developmental from exploitational cities are analyzed in Chapter 5.

NOTES

1. See United Nations, Department of International Economic and Social Affairs, *Patterns of Urban and Rural Population Growth* (New York: United Nations, 1980), Populations Studies 68.

2. See Kingsley Davis, *World Urbanization 1950-1970* (Berkeley: University of California, Institute of International Studies, 1969); United Nations, *Patterns of Urban and Rural Population Growth*, pp. 5-7.

3. For a detailed discussion, see Dennis A. Rondinelli and Kenneth Ruddle, *Urbanization and Rural Development: Spatial Policy for Equitable Growth* (New York: Praeger, 1978).

4. Gerald Breese, *Urbanization in Newly Developing Countries* (Englewood Cliffs, NJ: Prentice-Hall, 1966), p. 102.

5. Lim Heng Kow, *The Evolution of the Urban System in Malaya* (Kuala Lumpur: Penerbit Universati Malaya, 1978), p. 206.

6. Harm deBlij, *Mombasa: An African City* (Evanston, IL: Northwestern University Press, 1968), pp. 20-21.

7. Ibid., pp. 39-40.

8. Richard I. Lawless and Gerald H. Blake, *Tlemcen: Continuity and Change in an Algerian Islamic Town* (London: Bowker, 1976), pp. 11-12.

9. See A. A. Srivastaba, "Growth, Morphology and Ethnic Character in Ranchi-Dhurwa Urban Complex," in *Urbanization in Developing Countries,* ed. S. M. Alam and V. V. Pokshishevsky (Hyderabad: Osmania University Press, 1976), pp. 513-543.

10. Chakrit Noranitipadungkarn and A. Clarke Hagensick, *Modernizing Chiengmai: A Study of Community Elites in Urban Development* (Bangkok: National Institute of Development Administration, 1973), pp. 6-7.

11. Clifton W. Pannell, *T'ai-Chung, T'ai-Wan: Structure and Function,* Research Paper 144 (Chicago: University of Chicago, Department of Geography, 1973), p. 175.

12. William R. Hance, *Population Migration and Urbanization in Africa* (New York: Columbia University Press, 1970).

13. David W. Dent, "Urban Development and Governmental Response: The Case of Medellín," in *Metropolitan Latin America: The Challenge and the Response,* ed. W. A. Cornelius and R. V. Kemper (Beverly Hills, CA: Sage, 1978), p. 132.

14. John Gulick, *Tripoli: A Modern Arab City* (Cambridge, MA: Harvard University Press, 1967).

15. Breese, *Urbanization in Newly Developing Countries,* p. 66.

16. H. I. Ajaegbu, *Urban and Rural Development in Nigeria* (London: Heinemann, 1976), p. 41.

17. See Hance, *Population Migration and Urbanization in Africa.*

18. Bryan Roberts, "The Social History of a Provincial Town: Huancayo, 1890-1972," in *Social and Economic Change in Modern Peru,* ed. R. Miller, C. T. Smith, and J. Fisher (Liverpool: University of Liverpool, Center for Latin American Studies, 1976), pp. 136-197.

19. Srivastaba, "Growth, Morphology and Ethnic Character."

20. Gulick, *Tripoli: A Modern Arab City,* p. 35.

21. Janet Abu-Lughod, "Problems and Policy Implications of Middle Eastern Urbanization," in *Studies on Development Problems in Selected Countries of the Middle East,* ed. United Nations (New York: United Nations, 1973), pp. 42-62.

22. Walter F. Weiker, *Decentralizing Government in Modernizing Nations: Growth Center Potential of Turkish Provincial Cities* (Beverly Hills, CA: Sage, 1972), p. 65.

23. Ruben E. Reina, *Parana: Social Boundaries in an Argentine City* (Austin: University of Texas Press, 1973), pp. 27-28.

24. Hance, *Population Migration and Urbanization in Africa.*

25. Rhoads Murphy, "Urbanization in Asia," in *The City in Newly Developing Countries,* ed. Gerald Breese (Englewood Cliffs, NJ: Prentice-Hall, 1969), p. 61.

26. See Noranitipadungkarn and Hagensick, *Modernizing Chiengmai.*

27. Sen-Dou Chang, "The Changing System of Chinese Cities," *Annals of the Association of American Geographers* 66 (September 1976): 404.

28. See Kow, *Evolution of the Urban System.*

29. Pannell, *T'ai-Chung, T'ai-Wan.*

30. Robert A. Hackenberg and Beverly H. Hackenberg, "Secondary Development and Anticipatory Urbanization in Davao, Mindanao," *Pacific Viewpoint* 12, no. 1 (1971): 1-20.

31. Richard M. Morse, "Some Characteristics of Latin American Urban History," *American Historical Review* 67, no. 2 (1962): 332.

32. Richard M. Morse, "Recent Research on Latin American Urbanization: A Selective Survey with Commentary," in *The City in Newly Developing Countries,* ed. Gerald Breese (Englewood Cliffs, NJ: Prentice-Hall, 1969), p. 477.

33. Gulick, *Tripoli: A Modern Arab City.*

34. Lawless and Blake, *Tlemcen: Continuity and Change,* p. 83.

35. See E. B. Martin, *The History of Malindi* (Nairobi: East African Literature Bureau, 1973).

36. See deBlij, *Mombasa: An African City.*

37. Gulick, *Tripoli: A Modern Arab City,* pp. 32-33.

38. See Akin L. Mabogunje, "The Urban Situation in Nigeria," in *Patterns of Urbanization: Comparative Country Studies,* ed. S. Goldstein and D. F. Sly (Liege, Belgium: International Union for Statistical Study of Population, 1977), pp. 569-641.

39. A. G. Onokerhoraye, "The Spatial Aspects of Urban Growth in Nigeria: Some Planning Implications for National Development," *Cultures et Developpment* 8 (1976): 237-303.

40. V.E.A. Akorede, "The Impact of Socio-Cultural Changes on the Pattern of Urban Land Use: The Case of Ilorin," *African Urban Studies* 5 (Fall 1979): 71-84.

41. Josef Gugler and William G. Flanagan, *Urbanization and Social Change in West Africa* (London: Cambridge University Press, 1978), pp. 27-28.

42. Roberts, "Social History of a Provincial Town," p. 150.

43. See Arthur D. Murphy and Henry A. Selby, "Poverty and the Domestic Life Cycle in an Intermediate City of Mexico" (Paper prepared for the Workshop on Intermediate Cities, East-West Center, Population Institute, Honolulu, 1980).

44. Leighton W. Hazelhurst, "The Middle-Range City in India," *Asian Survey* 8, no. 7 (1968): 539-552.

45. Noranitipadungkarn and Hagensick, *Modernizing Chiengmai,* pp. 15-16.

46. Hackenberg and Hackenberg, "Secondary Development and Anticipatory Urbanization," p. 6.

47. Roberts, "Social History of a Provincial Town," p. 148.

48. Ibid., pp. 148-149.

49. Ibid., pp. 147-150.

50. Louis H. Muench, "Town Planning and the Social System," in *African Urban Development,* ed. M. Koll (Dusseldorf: Bertelsmann Universtätsverlag, 1972), p. 53.

51. Lawless and Blake, *Tlemcen: Continuity and Change,* pp. 149-150.

52. Ibid., p. 113.

53. Gulick, *Tripoli: A Modern Arab City.*

54. Srivastaba, "Growth, Morphology and Ethnic Character," pp. 517-518.

55. Arch Dotson and Henry Teune, "On the Consequences of Urbanization: Contribution to Administrative Capacity and Development," in *Metropolitan Growth: Public Policy for South and Southeast Asia,* ed. L. Jakobson and V. Prakash (New York: John Wiley, 1974), p. 34.

56. See Charles E. Chapman, *Colonial Hispanic America: A History* (New York: Macmillan, 1933), pp. 149-150; Lewis Hanke, *The Imperial City of Potosi: An Unwritten Chapter in the History of Spanish America* (The Hague: Marinus Nijhoff, 1956).

57. See Walter B. Stöhr, "Some Hypotheses on the Role of Secondary Growth Centers as Agents for Spatial Transmission of Development in Newly Developing Countries: The Case of Latin America," in *Proceedings of the Commission on Regional Aspects of Development of the International Geographical Union,* vol. 2, ed. F. Helleiner and W. Stöhr (Ontario: International Geographical Union, 1974), pp. 75-111.

58. Kow, *Evolution of the Urban System,* p. xxiii.

59. Wilbur R. Thompson, *A Preface to Urban Economics* (Baltimore: Johns Hopkins University Press, 1965), pp. 21-24.

CHAPTER 4

THE FUNCTIONS OF SECONDARY CITIES

A review of the relatively few studies that have been done of contemporary secondary cities in developing countries can provide insights into the kinds of functions they perform and can help to identify activities that can be promoted or strengthened in smaller towns. Only a few countries have detailed data on socioeconomic characteristics of all of their cities over 100,000 population; comparative analyses must, therefore, depend primarily on case studies and case histories of individual cities. This chapter draws most heavily on such cases in 17 developing nations in Africa, South America, the Middle East, and Asia. The cities for which case histories or case studies were used are listed in Table 4.1. The cities ranged in size from a little under 100,000 in population (Dagupan, the Philippines) to a little more than 2.5 million (Guadalajara, Mexico) in 1980. About half of these cities grew from a population of 100,000 or less in 1950; the others had reached that size earlier in the twentieth century. The paucity of data on systems of secondary cities in developing countries leaves only a few countries, such as South Korea, from which generalizations can be made about changes in secondary cities over time.

The case studies reviewed here offer some evidence that secondary urban centers can perform many of the functions described in Chapter 1 and that they can potentially contrib-

TABLE 4.1 Major Case History or Case Study Cities

Region and Country	City	Population 1980 (estimated)[a]	1970	1950	Major Sources
East and Southeast Asia					
Taiwan	Taichung	645,000	439,000	189,000	Pannell
South Korea	Daegu	1,579,000	1,050,000	389,000	Lee and Barringer; Ministry of Home Affairs statistics
Indonesia	Surabaya	2,440,000	1,517,000	613,000	Atmodirono and Osborn
	Semarant	880,000	645,000	378,000	Atmodirono and Osborn
	Pedang	281,000	194,000	161,000	Atmodirono and Osborn
	Medan	886,000	631,000	347,000	Atmodirono and Osborn
	Ujung Pandang	536,000	446,000	374,000	Atmodirono and Osborn
Malaysia	Penang-George-town	314,000	272,000	202,000	Osborn
Philippines	Davao	703,000	404,000	128,000	Hackenberg and Hackenberg
	Iligan	192,000	108,000	>100,000	Ulack
	Dagupan	c.100,000	>100,000	>100,000	Dannhaeuser
Thailand	Chiangmai	c.125,000	>100,000	>100,000	Noranitipadungkarn and Hagensick; Temple et al.
India	Ranchi	476,000	361,000	230,000	Srivastaba
	Meerut	432,000	244,000	105,000	Sundaram

Africa and
the Middle East

Algeria	Tlemcen	230,000	121,000	>100,000	Lawless and Blake
Lebanon	Tripoli	240,000	183,000	107,000	Gulick
Turkey	Eskisehir	378,000	278,000	124,000	Weiker
	Kayseri	435,000	275,000	114,000	Weiker
Kenya	Mombasa	396,000	256,000	>100,000	deBlij
	Malindi	>100,000	>100,000	>100,000	Martin
Zambia	Kitwe	445,000	220,000	>100,000	Kay
Nigeria	Ibadan	970,000	725,000	432,000	Ajaegbu; Muench
	Ilorin	562,000	239,000	>100,000	Akorede
	Ilesha	159,000	143,000	117,000	Trager
	Kano	289,000	190,000	>100,000	Mortimore; Lubeck
Latin America					
Mexico	Guadalajara	2,762,000	1,565,000	415,000	Walton
	Oaxaca	125,000	>100,000	>100,000	Beals; Murphy and Selby
Argentina	Parana	150,000	130,000	>100,000	Reina
Colombia	Cali	1,606,000	954,000	288,000	Bromley
	Medellin	2,439,000	1,474,000	469,000	Dent
Peru	Huancayo	207,000	116,000	>100,000	Roberts

a. Population counts taken from United Nations, *Patterns of Urban and Rural Population Growth*, Population Studies 68 (New York: United Nations Department of International Economic and Social Affairs, 1980), Table 48.

ute to balancing the pattern of urbanization and to stimulating the economies of rural areas. However, the nature of the analysis does not allow universal generalizations to be drawn about the potential of secondary cities in all countries. The cases show that these cities perform social and economic functions that are important for regional and national development in many countries, but that some of them perform some functions poorly and others within the same country do not perform them at all. The cases identify some of the common characteristics of secondary cities in different parts of the world, but they provide less information on the unique differences among cities in different cultures and geographical regions. Although a review of what seem to be common functional characteristics of many secondary cities can be useful as a framework for planning and in the formulation of development strategies, the generalizations that follow must be considered hypotheses or propositions about secondary cities that must be tested in each developing country before policies are formulated and programs are designed. What follows should be considered only a starting point for further research and analysis rather than a firm set of conclusions about the roles of all secondary cities.

The case histories and case studies of the 31 cities listed in Table 4.1 indicate that these urban centers can perform some or all of the following functions:

(1) They can provide convenient locations for decentralizing public services through municipal governments, field offices of national ministries or agencies, or regional or provincial government offices, thereby creating greater access for both urban and rural residents to public services and facilities that require population thresholds of 100,000 or more.

(2) They can offer sufficiently large populations and economies of scale to allow the concentration within them of health, education, welfare, and other services, and can act as regional or provincial centers for a variety of basic social services and facilities.

(3) They usually offer a wide variety of consumer goods and commercial and personal services through small-scale enterprises and through extensive "informal sector" activities.

(4) Many act as regional marketing centers offering a wide variety of distribution, transfer, storage, brokerage, credit, and financial services through their regularly scheduled and institutionalized markets or through periodic markets and bazaars.

(5) They often provide conditions that are conducive to the growth of small- and medium-scale manufacturing and artisan and cottage industries that can serve local markets and satisfy internal demand for low-cost manufactured goods, and some also support large-scale industrial activities.

(6) Many act as agroprocessing and agricultural supply centers for their regions and provide services to rural populations in their hinterlands.

(7) They often create conditions—through relatively high levels of population concentration, advantageous locations, marketing and agroprocessing functions, and linkages to rural communities—that are conducive to the commercialization of agriculture and to increasing agricultural productivity and income in their immediately surrounding hinterlands.

(8) They can be sources of off-farm employment and supplementary income for rural people and, through remittances of migrants, provide additional sources of income to people living in rural towns and villages in their regions.

(9) They often serve as areawide or regional centers of transportation and communications, linking their residents and those of rural villages and towns in their hinterlands to larger cities and other regions in the country.

(10) They can absorb substantial numbers of people migrating from rural areas to urban centers, transforming a "rural-to-primate city" migration pattern to a "stepwise" pattern, and offer long-term or permanent residence to some migrants, thereby creating a more balanced distribution of urban population.

(11) They can function effectively as centers of social transformation by: (a) accommodating social heterogeneity and encouraging the integration of people from diverse social, ethnic, religious, and tribal groups; (b) accommodating organizations

that help to socialize and assimilate rural people into city life, supporting them during their transition and mediating conflicts among them; (c) infusing new attitudes, behavior, and lifestyles that are more conducive to urban living; (d) providing opportunities for economic and social mobility; and (e) offering new economic and social opportunities for women.

(12) They can be channels for the diffusion of innovation and change, the spread of benefits of urban development, the stimulation of rural economies, and the integration of urban centers and rural settlements within their regions through social, economic, and administrative linkages.

As noted in Chapter 3, however, and as will be seen later, these potential functions of secondary cities do not come automatically with population growth or economic diversification. The kinds of economic activities that are established in secondary cities, their mix and combination, the organization and structure of their economies, and the nature of the development policies pursued within them by local governments, national agencies, and private business all affect the degree to which they are catalysts for development within their regions. Certain conditions, or preconditions, some of which were noted in Chapter 3 and others that will be described later, seem to be strongly associated with cities that have been more "developmental." The case studies and case histories reviewed here also identify an impressive set of problems that accompany secondary city growth and diversification with which governments in developing nations must be prepared to cope if they wish to develop such cities. These problems will be identified and described in more detail in Chapter 6.

Public and Social Service Centers

Many of the case histories of secondary cities point out that they have sufficiently large populations to offer economies of scale for a wide variety of health, education, and welfare services. Weiker notes in his study of the Turkish city of Eskisehir, for instance, that it grew into an important educational center for its region and supported a sizable number of middle and secondary schools, teacher training colleges, foreign language institutes, and vocational colleges and academies.[1] In Lebanon, a notable function of Tripoli was that it became, quite early in its history, one of the major centers through which religious organizations and the government extended social services to its region. Since the 1950s the contribution of private and religious organizations to education and social welfare has far surpassed that of government. Religious and charitable organizations in Tripoli offer a large number of social services: They run orphanages, provide food and clothing to needy families, operate girls' schools, pay funeral expenses for poor families, supply books and educational materials to students whose families cannot afford to buy them, provide scholarships, assist the poor with medical and health care expenses, and operate day nurseries for children of working mothers. Other organizations operate free primary and adult educational programs for illiterates, assist welfare patients in Muslim hospitals, provide vocational education for girls from poor families, operate homes for the infirm and aged, and provide girls' recreational, health, and special education programs.[2]

Similarly, Chiangmai emerged as an important center of public health care and education in Northern Thailand before it reached 100,000 in population. By the early 1970s, Chiangmai had over 900 primary and secondary schools, a major university, over 30 colleges, and vocational or special education schools, some run by the government and others by private organizations. Their importance to the development of the city and its region is only hinted at by the observation of Noranitipadungkarn and Hagensick that the "educational institutions instill new skills and values and open opportunities for a large number of youth, not only for local residents but also for those from other parts of the country."[3] At a time when the city's population was just over 100,000, it was able to support 2 of the best hospitals in the northern region, several smaller hospitals, numerous private clinics and dispensaries, a malaria eradication clinic, a tuberculosis center, and other special-care facilities. Chiangmai's provision of health care services to the northern region became extensive enough in the early 1970s for the system to treat many who previously had to go to Bangkok for treatment.[4] Moreover, the facilities attracted new residents to Chiangmai, including "professionals who in turn are likely to be supporters of additional urban growth in activities such as education, health, engineering, commerce and the arts."[5]

That secondary cities in Indonesia are important social and educational service centers for their regions is reflected in the fact that by the early 1970s Surabaya had more than 27 percent of the senior high schools in East Java, half of all elementary schools for the mentally retarded, and more than 30 percent of senior and junior high school students, although the city's population was only 6 percent of the region's total. Medan had 87 percent of northern Sumatra's high school and college students and 60 percent of the maternity clinics in the region. Ujung Pandang (Makassar) had 96 percent of South Sulawesi's university and college students, nearly 40 percent of the senior high schools, more than 40 percent of the health clinics, and a quarter of the regions's hospitals. The city's population was less than 10 percent of the region's. Pedang, with about 7 percent of West Sumatra's population, had 95 percent of the region's college students, more than 40 percent of the senior and 30 percent of the junior high school students, and 85 percent of the region's paramedics.[6]

As studies of Iligan City in the Philippines have shown, and as other studies that will be cited later confirm, secondary cities that are industrializing tend to attract migrants who are likely to demand more diversified and higher-quality social services. The migrants to Iligan during its period of industrialization in the 1970s had higher levels of educational attainment than the national average—migrants working in the city's industries had twice the level of educational attainment of the population over 20 years old in the Philippines, and they in turn created demand for better educational services for their children.[7]

However, Yu's studies of educational services in Taegu, Korea's third largest metropolitan area, found that this secondary manufacturing and commercial city became an educational center for its region primarily because of the economies of scale created by the concentration of population, and that "the structure and functions of education in Taegu and Taegu's industrialization have not had much direct effect on each other." Although industrialization undoubtedly attracted many of the people who came to Taegu, Yu claims that "it has been urbanization—the concentration of population in Taegu—which has much more effect and bestowed more functional benefits on education in Taegu." And, he insists, "the opportunities of Taegu's educational system have had a strong reciprocal effect on the process of urbanization." As the city became a center of educational services in the southeastern region of Korea, it tended to attract people interested in obtaining better education for themselves or their children than they could get in smaller cities or rural villages.[8]

Statistics on services in Korean secondary cities indicate that even those just over a 100,000 in population can support at least basic educational and medical programs and that the number, diversity, and quality of health and educational services tend to be higher in larger secondary cities. In 1978, the 30 cities with more than 100,000 people—excluding Seoul—accounted for 40 percent of the medical personnel—doctors, nurses, and dentists—located in urban areas and 42 percent of all medical facilities in urban areas. In 1975, Korean secondary cities had nearly 44 percent of the elementary, middle, and high schools. Although the mix of personnel and types of facilities differed among cities in different size categories—the larger cities tended to have a greater percentage of licensed doctors among their medical personnel, a larger number of general and specialized hospitals among their medical facilities, and a larger percentage of high schools in their share of educational facilities—in most cases the share of educational and medical facilities in secondary cities was closely related to their share of urban population (see Table 4.2).

Seoul still has a high percentage of the health and educational services within the country, both because of its relatively high concentration of urban population and because of the economies of scale and other benefits it offers to specialized professionals and institutions. Nearly half of the medical personnel and facilities in urban areas remain concentrated in the 5 largest secondary cities, but the share in smaller secondary cities is only slightly smaller than their share of urban population: Cities of from 100,000 to 500,000 in population had a slightly higher share of schools than that of their urban population. Although most of the largest, most prestigious, colleges and universities remain in Seoul, 16 of the secondary cities were able to support 118 colleges, universities, and other institutions of higher learning in 1975. The government's policy of requiring universities in Seoul to expand by establishing branches in intermediate cities will enhance the function of secondary cities as educational centers in the future. There is some evidence from Korea, however, that as secondary cities grew during the 1960s and 1970s, the expansion of health and educational services did not keep pace with population increases. Although services tend to be well distributed in Korean secondary cities, the numbers of medical personnel, health facilities, and schools per 1,000 population has decreased in all city-size categories.[9]

TABLE 4.2 Distribution of Social and Public Services Among Secondary Cities of Various Size Categories in Korea

Population Size Category 1978	Number of Cities 1978	% Distribution of					% of Area Served by Piped Water 1978
		Medical Personnel 1978[a]	Medical Facilities 1978[b]	Schools 1975[c]	Universities and Colleges 1975	Teachers 1975	
National capital	1	50.2	42.2	17.8	23.4	51.7	92.0
500,000 or more population	5	24.7	26.6	20.5	41.5	30.6	36.6
499,999 – 200,000	7	7.3	6.6	8.9	17.5	7.6	81.5
199,999 – 100,000	18	7.5	8.9	14.1	17.5	9.6	67.4

SOURCE: Calculated from Republic of Korea, [Long Range Planning of Urban Growth to the Year 2000: Data Collection], vols. 1, 2 (unofficial trans.) (Seoul: Ministry of Construction, 1980).

a. Includes licensed medical doctors, dentists, and nurses.
b. Includes hospitals and clinics.
c. Includes elementary, middle, and high schools.

This is a common characteristic of secondary cities in many developing countries. Public services generally have been increasing in secondary cities, although not always as fast as in the national capital or at pace with population growth. In Colombia, for instance, governments in secondary cities have taken on increasing responsibilities for providing utilities, infrastructure, and social services, and the national government has also been expanding public services in intermediate urban centers. Local governments in Cali and Cartegena, for instance, now have responsibility for water and sewerage services, as do 79 other municipalities. They and many other municipalities also have responsibility for local telephone service, roads and bridges, street lighting, refuse collection, parks and recreation, fire protection, markets, and cemeteries, and share some responsibility with national ministries in providing health, social welfare, and housing services. During the 1970s, municipal government expenditures on services in secondary cities contributed from 17 to 50 percent of all public sector spending in Colombia.[10] Moreover, the coverage of urban services has been increasing: In cities of 100,000 or more population about 77 percent of the residents were served by public water systems by the early 1970s, and nearly 60 percent were served by sanitary sewers, with many individual cities achieving substantially higher levels of coverage. World Bank studies indicate, however, that service coverage was not only highly variable *among* secondary cities, but also *within* them. The poorest residential areas of Colombian secondary cities tended to have the least access to health, educational, water, sewerage, or other urban services, and thus did not receive "the threshold benefits associated with access to public services, which is likely to bias the incidence of public expenditures towards the regressive side."[11]

Secondary cities in Asia also seem to be able to support relatively high levels of public utilities. Almost all of the

larger secondary cities in Indonesia receive nearly uninter-
rupted electrical power in homes as well as in businesses and
public facilities.[12] By 1978, piped water had been extended
to a large portion of secondary cities in Korea; in cities of
500,000 or more residents about 87 percent of the city was
served by piped water, and some larger regional centers such
as Gwangju and Taegu had nearly total coverage (see Table
4.2). Coverage tended to decline in smaller cities, although
not in direct proportion to population size. About 81 percent
of the area was supplied with piped water in cities of from
200,000 to 500,000 population, and 67 percent coverage had
been attained in cities with from 100,000 to 200,000
people.[13]

Although in many countries the number, coverage, and
quality of services in secondary cities are often far below
those of the national capital, many secondary centers are able
not only to support a wide variety of basic public and social
services and facilities, but to provide greater access to these
urban functions to people living in their rural hinterlands.

Commercial and Personal Service Centers

Although economists often regard the "tertiary" sector of
urban economies as parasitic rather than productive and cities
with high levels of employment in commerce and services as
"overurbanized," it is becoming clear that as cities grow, in
both Western and developing nations, the commercial and
service sectors become an increasingly important source of
income and employment for urban residents. In much of the
developing world, cities have primarily commercial and
service—rather than manufacturing—economies. As noted in
Chapter 2, in the early stages of urbanization the economies
of secondary cities are usually dominated by basic consumption and personal services; as they grow, manufacturing and
the productive services—construction, transport, communications, financial, repair, and related activities—tend to become

more important, along with wholesaling and retailing of durable consumer goods.

EMPLOYMENT IN COMMERCE AND SERVICES

In many cities, the tertiary sector has been, and will continue to be, the base of the economy. Surveys of the employment structure in Mexican secondary cities indicate that commerce and services consistently absorbed from 40 to 45 percent of the workers in cities over 100,000 population during the 1960s and 1970s and that construction, utilities, and transportation accounted for an additional 10 to 12 percent of employment. Only about 15 to 20 percent of the labor force was engaged in manufacturing. Even in the large secondary cities in Mexico, commerce and services dominate the employment structure.[14] In Guadalajara, a highly industrialized metropolis, more than half the labor force worked in the tertiary sector, although the structure of commerce has been changing since the 1950s, with increased shares of the market going to chain discount and branch stores of larger enterprises that now compete more effectively with the small business and neighborhood stores that made up the tertiary sector before the 1960s.[15] In Monterrey, Mexico's third largest city, more than half the male population was engaged in commerce or services in the 1960s and 1970s; nearly 10 percent worked in construction, more than 18 percent were employed in transport, communications, and public enterprises, and 11 percent in other services. In 1970 only a third of the labor force in Monterrey was employed in manufacturing.[16] Similarly, in one of Mexico's smallest intermediate cities, Oaxaca, nearly two-thirds of the labor force was engaged in commerce or services in 1970, which was typical of the percentage employed in the tertiary sector in cities with populations of less than 400,000.[17]

Research on regional development in Mexico indicates that secondary cities, especially those in rural regions, tend to have high concentrations of commercial activities and that their strong role as regional commercial and service centers is attributable in part to the lack of settlements with sufficiently high levels of population concentration to support commercial enterprises. In a sense, the largest cities in rural regions, for whatever reasons they grew, tend to capture the market for commercial and personal services; the paucity of smaller cities reinforces their positions and strengthens their advantages as commercial centers.[18] Studies of cities in the West Central region of Brazil also indicate that they have a high percentage of people employed in the services sector and in wholesale and retail establishments and that in nearly all smaller towns the tertiary sector dominates the economy.[19]

In the Middle East, Gulick reports that the lifeblood of Tripoli's economy runs through thousands of small shops concentrated in the suqs (markets) and along narrow back streets of the city. The shops make or sell cloth, woodwork, leather goods, shoes, metal utensils, jewelry, tools, pots, trays, and an endless variety of small consumer goods. They are operated on a small scale and in traditional fashion, but seem adaptable to changing conditions.[20] This diversity of commercial activities appears in Middle Eastern cities and towns before they reach intermediate size. In Kashan, Iran, a city of about 60,000 people in the early 1970s, Costello surveyed over 2600 retail, wholesale, and service establishments, including more than 140 bakeries and confection shops, 106 shoemakers or shoe shops, 113 carpenter shops, more than 100 metalsmiths, 60 motor and bicycle repair shops, 600 groceries and butchers, over 100 tailors and sewing shops, and as many clothing and yarn stores. The city had more than 400 transport and storage establishments, 9 small hotels and guest houses, 20 photographers, and assorted other recreational and personal services. All were small-scale operations employing 1 or 2 people, usually the owner and a family member.[21]

Commerce and services also continue to be the foundation of the economies of most secondary cities in East and South Asia. Atmodirono and Osborn report that 80 percent of the labor force in Surabaya, Indonesia, was employed in commerce, professional and personal services, transportation, and construction in the early 1970s, and that only about 15 percent was employed in manufacturing. More than two-thirds of Medan's workers were engaged in trade and commerce, and an additional 15 percent were employed in transport, communications, and services.[22]

Studies of cities of 100,000 or more residents in India showed a strong correlation in the 1950s between population growth and "change in the total and nonagricultural labor force and employment in the tertiary, rather than the secondary, sector."[23] As Ahmedabad grew in population during the 1960s, for instance, there were noticeable shifts in its employment structure: Between 1961 and 1971, employment in manufacturing declined by almost 7 percent and the labor force engaged in commerce and trade increased by more than 4 percent, and in transport by nearly 2 percent.[24]

Sundaram postulates from his study of Meerut City, in India, that secondary cities in that country may begin to shift from basic services to production-oriented commercial activities and services, along the lines suggested in Chapter 2, when they reach a population of about 300,000. Commercial and service establishments catering to industrial needs tend to become more dominant as the city grows beyond 300,000 in population; employment in transport, storage, construction, communications, wholesale trade, finance, insurance, real estate, and other production services increases, and employment in basic consumer goods and services either stabilizes or declines slightly. Sundaram concludes that "perhaps medium size cities undergo a critical phase of transformation in their urban economic structure while crossing a population threshold around this range."[25]

Some indication of the changes that occur in the economies and employment structures of secondary cities as a developing nation urbanizes and industrializes comes from South Korea. In 1980, commerce and services still dominated the employment structure of most secondary cities, but their economies changed quite drastically over 20 years. In 1960, more than half of the labor force in all secondary cities was employed in commerce and services, with services alone accounting for nearly a third of secondary city employment. Wholesale and retail trade establishments employed about 20 percent of the labor force in large secondary cities—those with over 500,000 population—and for a little more than 17 percent in cities with between 100,000 and 500,000 residents. The smaller cities still had substantial percentages of their labor force engaged in agriculture.

By 1980, commerce and services still played a strong role in the economies of secondary cities, accounting for about 44 percent of employment in those with more than 500,000 people, nearly 40 percent in cities with populations of between 200,000 and 500,000, and nearly 56 percent in cities of from 100,000 to 200,000 people, but the percentage of the labor force engaged in social and personal services dropped drastically in all population size categories between 1960 and 1980 and to less than 10 percent in the largest cities, about 13 percent in cities of between 200,000 and 500,000 people, and about 17 percent in smaller intermediate cities (see Table 4.3).

Employment in wholesale and retail trade increased in the largest and smallest secondary cities, as did employment in production-oriented services, and both declined slightly in cities of from 200,000 to 500,000, where the increase in manufacturing employment was greatest. Manufacturing came to dominate the employment structure of cities with populations over 200,000, and in all secondary cities, employment in the primary sector declined drastically, to no

TABLE 4.3 Distribution of Employment by Sectors in Secondary Cities of Different Size Categories, South Korea: 1960, 1974, 1980

Population Size Category 1978	Number of Cities	% Distribution of Employment														
		Agriculture and Mining			Manufacturing			Wholesale and Retail Trade			Construction, Utilities, Transportation, Communications			Services		
		1960	1974	1980	1960	1974	1980	1960	1974	1980	1960	1974	1980	1960	1974	1980
500,000 or more	5	14.5	6.3	1.4	21.3	30.4	55.6	19.9	27.8	21.7	11.2	15.3	12.6	32.0	19.8	9.6
499,999 – 200,000	7	17.4	14.2	6.6	16.9	28.7	55.2	17.4	22.1	16.5	10.4	14.2	8.8	34.0	20.5	13.2
199,999 – 100,000	18	28.1	20.4	4.7	14.1	21.3	39.7	17.5	23.4	19.2	8.8	12.9	9.2	31.0	21.4	17.1

SOURCES: Compiled from Republic of Korea, [Long Range Planning for Urban Growth to the Year 2000: Data Collection], vols. 1, 2 (unofficial trans.) (Seoul: Ministry of Construction, 1980); and Republic of Korea, *Municipal Yearbook of Korea, 1980* (Seoul: Ministry of Home Affairs, 1980).

more than 7 percent of the labor force in any city-size category. Thus although the tertiary sector is no longer predominant in the largest Korean secondary cities, it still dominates the economies of smaller ones and is an important source of employment and income in them all. In 1980 urban commercial establishments provided employment for nearly 45,000 people, and production-oriented services offered 197,000 jobs. Social and commercial services employed more than 280,000 workers.

Even in relatively large secondary cities, the tertiary sector tends to be composed of large numbers of small establishments. In 1968, during a period of rapid growth and transformation in Korea's third largest city, 99 percent of the nearly 21,000 stores in Taegu had less than 4 employees. The characteristics of the commercial sector there appear frequently in case studies of secondary cities in other parts of the world: The businesses operate in places where family income is not high and thus must cater to demand for small purchases of consumable goods. Owners have only small amounts of capital to invest in inventory. Businesses are run according to traditional methods, with informal accounting and management procedures. They usually employ only family members or close relatives, they survive on a small margin of profit, and their owners receive relatively low levels of income.[26]

As cities in Korea grew in population, the average size of commercial and service establishments increased slightly, but the tertiary sector is still composed primarily of small-scale enterprises. In all secondary cities except Pusan, Taegu, and Gangneung, the average wholesale and retail establishment employed only 2 people in 1980. Service establishments in cities with populations of more than 500,000 employed slightly larger numbers of workers, but in most cities of between 100,000 and 500,000 residents, services remained small businesses. In only a few cities did the average number of employees exceed 7.

INFORMAL SECTOR EMPLOYMENT

Employment statistics tend to underestimate the strength of the tertiary sector in developing countries: In most secondary cities a substantial number of the poor, recent migrants, women, and children earn their income through casual work or through trading in the "informal" sector. Even the biggest metropolitan centers usually have large informal sectors—composed of street vendors, traders, hawkers, and odd-jobbers—that support large numbers of people, albeit sometimes at or near subsistence levels. Bromley argues that "street trading is one of the most visible and important occupations in most African, Asian, Middle Eastern and Latin American cities."[27]

Bromley's studies of Cali, Colombia, a secondary city of more than a million people, found nearly 10,000 street sellers of food, manufactured goods, and newspapers alone, as well as substantial numbers of people engaged in selling lottery tickets, offering personal services such as shoe shining and repair, transporting small items, standing in queues for tickets, and other such occupations. Most of those selling food, manufactured goods, and newspapers have fixed locations, work without assistants, and concentrate themselves around places that attract large numbers of people—the municipal markets, stadiums, entertainment centers, and transport terminals—although street traders can be found throughout the city. About 80 percent of those surveyed by Bromley work without trading licenses, but only a small percentage are involved in petty crime or vice. About two-thirds are male, although women traders are often found in the markets and bazaars. Most are migrants who have lived in the city for more than five years and have little or no formal education.[28]

Bromley notes that in Cali, as in many other cities in developing countries, street traders are often considered nuisances, parasites, beggars, or thieves by officials and the elite, and they are often openly harassed by police and officials.

Yet Bromley and others have found that only a small minority of street traders are engaged in crime or begging and that participants in informal tertiary services and trade play an important role in providing low-cost goods to the poor, in obtaining at least a subsistence income for themselves and their families, and in providing "an important distribution system for many Colombian importers and manufacturers and for several multinational companies."[29]

Studies of African secondary cities also note the high level of participation in informal sector activities. In West African secondary cities the informal sector provides employment for large numbers of men and women and is an important source of training and apprenticeship in artisanal production and small-scale manufacturing and service, including carpentry, tailoring, shoemaking and repair, metalworking, printing, weaving, and small-machinery repair.[30] In Kumasi, Ghana, the informal sector provides employment for from 60 to 70 percent of the city's labor force, and 90 percent of the workers employed in the informal sector received their training or apprenticeship by working with others in the informal sector.[31] Thus for most secondary cities in developing countries the small-scale commercial and service sector and the informal tertiary sector provide an important source of employment and income for urban residents and are a crucial part of the urban economy.

Regional Marketing and Trade Centers

Because of the strong role they play in offering services and supporting commercial activities, secondary cities often act as markets and trade centers for their regions. This has been a historical function that continues to be of crucial importance for regional and national development. Many secondary cities become channels for the bulking, sale, distribution, and trading of rural products and for the distribution of urban manufactured goods within rural towns and villages before they reach 100,000 in population. Those that grow to this size often do so, at least in part, because of their higher concentration and greater diversification of activities that depend on and support their regional markets. Noraniti-padungkarn and Hagensick note that Chiangmai became an important regional market for northern Thailand early in the twentieth century, but the introduction of modern production techniques and the strengthening of its transport and communications linkages in the 1960s reinforced and expanded its marketing function. The city now plays an important role as a distribution point for locally produced goods and those manufactured in Bangkok and for exporting local goods abroad. Fresh and processed agricultural products are brought from Chiangmai's rural areas every day to be sold or traded in the city's fresh food markets and grocery stores. "As a market center, Chiangmai attracts many people to procure its goods," these authors note, and these people include not only the city's own residents, but those from outlying districts as far away as the city of Lampang, college and university students in the area, members of various military units stationed in the region, and, occasionally, people from the hill tribes and villages of Burma.[32]

Regional markets—organized and operated by local or provincial governments or by private stallholders—are, of course, familiar physical features of nearly every secondary city in developing countries. In his study of India, Hazelhurst notes that markets in secondary urban centers tend to be organized and operated on a regular basis, usually meeting daily or several times a week. They consist of numerous small-scale operators, vendors, and merchants, who usually employ the labor of family members and occasionally of one or two assistants, and who operate according to traditional management and accounting procedures, establishing regular relationships with suppliers and customers.[33] In larger urban markets there may be fixed prices, although bargaining and haggling often continue; there is a stronger set of relationships between buyers and sellers over the long run than in smaller, rural, periodic markets; and vendors are more likely to extend credit to regular customers and to obtain credit from suppliers. Some vendors also act as intermediaries, brokers, and truckers as well as traders. The marketing system in middle-sized cities in India, and in many other developing countries, is usually open; whoever has the resources and entrepreneurial inclination can usually participate. Moreover, trading activities in the institutional market usually "spill over" into adjacent areas of the city, where vendors and hawkers who cannot afford stalls sell goods in small lots or offer services.

Hazelhurst observed the mechanisms that exist in the market systems of secondary cities to assimilate outsiders into their economic structure. He notes that outsiders usually must compete with local people to obtain the resources needed to operate effectively in local markets. Outsiders must also compete with local venders and stallholders for the potentially vast clientele from surrounding regions that the secondary cities serve.[34] The fact that these cities serve large regions is crucial to the maintenance of their dynamism and diversity. "The region provides a setting within which one can secure a position in the flow of economic life," Hazelhurst points out, "and, most important, secure a basis for legitimate creditor-debtor relationships with other merchants of the region."[35]

Even small secondary cities play important roles as regional marketing centers. Although "informal" trade activities dominate market systems in small secondary cities, they are often organized on at least two or three different levels, with complex networks of interaction among participants. Studies of Dagupan City, in the Philippines, for example, point out that it has served as an important trade and distribution center for the northern Luzon plain for more than 100 years, acting as a channel for the transfer of produce raised in its rural regions for the Manila markets, and for the distribution of manufactured goods from Manila and abroad within Dagupan's local service area.[36] The marketing systems in Dagupan facilitate the trading and distribution of produce, meat, and fish from rural producers in its hinterland to city consumers, and to groceries and packing companies in Manila. Traders also import packaged goods and manufactured items from the capital city.

Dagupan, with a little less than 100,000 residents, supports a complex, dualistic marketing and distribution system composed primarily of numerous small-scale vendors of groceries and cheap consumer goods, but with larger wholesale and processed food distribution establishments and brokers and agents of Manila companies also participating actively in the city's trade. According to Dannhaeuser, the wholesalers of food products—mainly Chinese and *mestizo* independent traders—purchase a wide range of goods from agents of large Manila companies and sell to vendors and grocery store operators in Dagupan. Small branch stores of companies based in Manila offer bulky and capital goods to shopkeepers and distributors in Dagupan. But the city's regional marketing functions depend mainly on hundreds of small-scale neighborhood stores and market-stall vendors who obtain their goods from both producers and wholesalers and who engage in direct exchange with farmers and city consumers. Dagupan's traders, like those in Indian middle-range cities, operate on a small scale; their businesses are conducted by members of their immediate families, with few or no employees, and their margins of profit are small. Even the largest wholesale establishments in the city were family owned and operated in the 1970s, and they made only about 500 to 600 transactions a day.[37] Studies of marketing systems in the Middle East, Latin America, and Africa come to similar conclusions about the importance of these activities in the economies of middle-sized cities and about the characteristics of secondary city marketing functions.

In the small secondary cities in the Middle East the *suq* and, in Latin America, the *plaza* play an important economic and social role. Beals's description of the influence of the market in Oaxaca, Mexico, could just as well describe the role of the plaza in small secondary cities in any country in Latin America:

Despite Oaxaca City's great development of modern commercial establishments and involvement in the distribution of industrial products, the Saturday *plaza* dominates the rhythm of activity, not only for most residents of the City but for people in the surrounding villages. While this is primarily true of economic activities, the influence of the *plaza* permeates many other aspects of daily life. In a market town the day of the *plaza* is the time when government offices are busiest, political activity is greatest and many churches have their greatest attendance. The *plaza* often affects the rhythm of household life. In the City the *plaza* day is the time when, even in upper-class households, the mistress (usually accompanied by a servant) may participate in large-scale family shopping, storekeepers do their largest volume of business, buses from the villages run more frequently and carry more passengers and pedestrian and automobile traffic are at a peak.[38]

The market in Oaxaca remains a vital base of the city's economy, as do markets in other cities in the developing world. The main urban market influences the process of village and town trade within its region, and acts as a channel of distribution for farm and handicraft products from rural areas and for manufactured goods from Mexico City and other large urban centers. It organizes a good deal of the small-scale trade within the city and provides much of the revenue for the city's budget.

Moreover, the market organizations in Oaxaca provide political influence for their members in local government decisions. The market channels credit and other financial resources within the city and provides employment and income for people in a variety of jobs. Indeed, the markets in secondary cities such as Oaxaca generate much of the commercial and service employment that dominates their economies. In Oaxaca, the market offers opportunities for small farmers and poor campesinos to trade their produce, vegetables, and grains; for stock raisers to sell cattle, goats, sheep, and dairy products; for gatherers to exchange firewood, lumber, charcoal, lime, and other uncultivated products; for fishermen to sell their catches; for artisans to sell or trade textiles, pottery, baskets, woven materials, iron, brick- or woodwork, household utensils, and an enormous variety of household goods. The markets offer opportunities for purveyors of services to reach mass consumers most efficiently. Carpenters, masons, bakers, butchers, barbers, midwives, blacksmiths, tailors, seamstresses, stonecutters, traditional healers, and marriage brokers ply their trades in or near the markets along with mechanics, repairmen, doctors, druggists, agricultural suppliers, and others whose permanent shops or offices are located near the plaza or in other areas of the city. The market offers an outlet for many goods produced in the city—shoes, textiles, ready-made clothing, furniture, soap, processed and packaged goods, flour, coffee, sugar, soft drinks, and bakery goods. Moreover, the market in Oaxaca supports traveling vendors, storekeepers, agents, brokers, middlemen, and truckers. It stimulates business for bus, truck, and taxi owners and creates opportunities for commission agents, money lenders, warehouse owners, and others who faciliate market trade.[39]

Although the markets of secondary cities often bring only subsistence or near-subsistence incomes for many participants, have limited new employment potential, and return only a small margin of profit to stallholders and vendors, they do engage the labor of large numbers of urban residents. And, as Dannhaeuser notes in his study of Dagupan City, not only has the marketing sector grown in many secondary cities in developing countries, but in places such as Dagupan "numerous cases exist in which third-order traders have been able over the course of one or two decades to establish substantial enterprises."[40] This produces a demonstration effect that stimulates and maintains entrepreneurial activities, for, as Dannhaeuser observes, "such successful careers are known and appreciated by the city's traders, which is part of the reason many individuals, especially those with some capital resources to start with, enter trade and maintain their positions in it even if success is not immediately forthcoming."[41]

Some studies from Africa and Asia suggest that as urbanization continues there is a gradual transformation of the marketing systems in secondary cities, a tendency for regional trade activities to shift from periodic and regular markets to stores and shops, although the traditional market may continue to be an important place of trade and commerce at which the poor, and bargain hunters, can obtain lower-priced goods. Studies of markets in secondary cities in Latin America indicate, moreover, that as population grows and densities increase in rural areas surrounding the cities, greater demand is generated in urban markets, trade within the hinterlands of these cities tends to become more concentrated in the larger, more diversified town markets, and the number of smaller periodic markets in surrounding villages tends to decline.[42]

Agroprocessing and Supply Centers

Studies of secondary cities in a number of developing countries point to their strong role as agricultural processing, supply, and service centers, and their function as centers of agroindustry and agribusiness. Although a relatively small percentage of the labor force in most cities is engaged directly in agricultural production, a substantial percentage of those employed in manufacturing and services may work for agroprocessing or commercial establishments serving farmers and the rural population.

The case histories of many cities with between 100,000 and 500,000 people show that they are highly dependent on their agricultural hinterlands, although agriculture may play a less important role in their economies as they grow and diversify. Gulick notes that Tripoli's economy depends on its agricultural hinterlands. The city has grown as an agroprocessing center and citrus orchards occupy much of the unbuilt land in and around Tripoli. Commercial establishments within the city export organges, lemons, olives, and other crops. In addition, confection industries, extractors, and manufacturers of sugared fruits and nuts have been an important part of the city's economy. Olive production provides inputs to the soap manufacturing industries located in Tripoli as well as to olive oil processors and residue fuel and fertilizer processing industries.[43]

Establishments in Chiangmai, Thailand, export agricultural and forest products, minerals, and crafts made from local wood and timber. Tobacco and timber have been especially important to the city's economy.[44] In 1977, the largest percentage of gross provincial product in Chiangmai—nearly one-third—was contributed by agricultural industries, with wholesale and retail trade contributing one-fourth. Strong contributions to gross regional product were also made by agriculture and agriculturally related businesses and industries in other secondary cities in Thailand.[45] Nearly 91 percent of Chiangmai's industries now consist of rice mills, wood product and craft establishments, and food processing firms. The largest industrial establishments in Chiangmai are the 76 tobacco companies, employing an average of 81 workers each.

Many of the secondary cities in India have grown, and continue to expand, as agroprocessing and service centers, and have diversified into other industries and services largely as a result of their agricultural processing and supply activities. Sundaram points out, in his study of Meerut City, that it "has grown mainly in response to the felt needs of its agricultural hinterland and it is the latter which has been feeding and sustaining its growth."[46] Industrial surveys show that the city's economy has depended heavily on sugar processing, the manufacture of agricultural implements, and commercial and service activities that supply the needs of a relatively prosperous agricultural population in its district, one in which nearly half the work force is still employed in cultivation. Although less than 3 percent of the city's labor force is directly engaged in cultivation, in the early 1970s nearly 19 percent were employed in food and food products industries, and nearly 17 percent more worked in machinery industries producing agricultural equipment and components. In addition, many of the city's service workers were employed in shops repairing agricultural implements and machinery.[47]

Their relatively large populations and their functions as regional markets and agroprocessing, service, and transport and distribution centers allow secondary cities in many developing countries to stimulate agricultural production in their hinterlands. Many of the case histories cited here remark on the intensification of agricultural production in the hinterlands as secondary cities grew, the tendency for farmers to switch from subsistence to cash-crop production, the introduction of new agricultural methods and procedures, and the diversification of village economies to meet the demands for agricultural and artisanal goods within the secondary cities.

Mortimore points out that as the Nigerian city of Kano grew, the "close-settled zone" around the city exhibited many of these characteristics: It became more densely populated, farmers shifted from subsistence to cash-crop production, land tenure became more individualized, land values increased, land subdivision became more frequent, and farming methods changed. Beginning in the 1960s, heavy manuring replaced fallowing, farmers began intercropping legumes and grains, and there was a marked increase in the hiring of agricultural labor, all of which intensified as commercial agriculture spread in the rural towns and villages in Kano's hinterland. Mortimore concludes that in the areas surrounding Kano, "the evidence suggests that a growing density of population is encouraging not only the intensification of agriculture, but also commercialization of the rural economy ... and the development of secondary sources of income dependent on markets."[48]

Studies of rural food processing industries in northern Nigeria report that growing demand in the large markets of the city of Zaria stimulated the production of *fura*—cooked balls of grain eaten with sour milk—by women in nearby rural villages. The women who made *fura* often employed other women in the village to process and prepare it and children and students to carry their goods to market. The increasing population and demand for consumer goods in Zaria created many opportunities for women from nearby villages to engage in trade—they acted as brokers between village enterprises and suppliers in the city. The expanding market in Zaria created demand for crafts, apparel, mats, and decorative household goods.[49] Although growing urban markets increased the demand for rural products in early stages of urban growth, the city also spawned competition for rural enterprises as it reached a size that allowed larger establishments to produce the same or substitute goods using more modern technology or mass production.

Evidence of similar effects of secondary city growth comes from Asia. Studies of changes in agricultural production in rural areas around Meerut City in India indicate a steady increase in both production and income as rapid changes occurred in farming areas surrounding the city in the 1950s and 1960s. The growing demand for agricultural products in secondary cities such as Meerut combined with the introduction of "Green Revolution" technology to bring a shift in cropping patterns from staple to cash or high-yielding varieties such as maize, wheat, and sugar cane. [50] In Indonesia, as well, many secondary cities have grown as the result of their agricultural processing and distribution functions. Surabaya, for example, is the major point of distribution and processing for nearly all agricultural products grown in its region that are not used for local subsistence.

Studies of Taegu, Korea, report similar changes in the rural areas surrounding that city in the 1960s as it grew into a regional commercial and manufacturing center. Park observes that agricultural resources in Taegu's hinterland came to be used more productively than those in remote areas; croplands in the areas around Taegu were producing more and generating higher income, farmers began to use hired labor more extensively, and members of farm families obtained off-season and part-time work in the city to supplement their incomes. Population growth in the Taegu area increased the demand for new agricultural products, including fruit, vegetables, livestock, poultry, grain, and tobacco. As demand increased, land was used more efficiently, the use of manufactured farm tools and implements spread rapidly, and the production of farm machinery and equipment became an important part of Taegu's economy.

Farmers living closer to the city improved their managerial abilities and tested new production and cultivation techniques more quickly than farmers living in areas farther from Taegu. Park concluded from his analysis that

the urbanization and industrialization of Taegu have a complementary relationship with the increasing commercialization of regional agriculture. In the Korean agricultural setting, the decentralization of urbanization and industrialization is an accelerating factor for modernization of the rural sector. And modernization of the surrounding rural sector contributes to the industrial and commercial growth of the city, by providing a market not only for more consumption goods, but, as farm incomes increase, also for more consumption goods wanted by farm people. With better roads, people get to town more often even though they would seldom go to Seoul.[51]

Some indication of the effects of urban growth on agriculture in Korea are reflected in changes in agricultural land use in secondary cities. Between 1967 and 1978, land in agricultural use increased from about 388 km² to about 446 km² in the 18 cities with populations of 100,000 and 200,000, but declined slightly in secondary cities larger than 200,000 population. As cities grew in population, up to about a quarter of a million, they seem to have generated increased demand and provided larger markets for agricultural products, and more land was brought into agricultural production in and around those cities. But as cities continued to grow to over 250,000 population, other employment opportunities were created, population densities on the peripheries and within the cities began to rise, land values increased and agricultural production became more efficient, and pressures were created for the conversion of agricultural land to other uses. Agricultural production in the hinterlands, however, could increase or be maintained with less land in and around the city devoted to agricultural use.[52]

Centers of Small-Scale Industry

Although some of the cities that have reached 100,000 or more in population in recent years—especially in Taiwan, Korea, India, Nigeria, and Brazil—grew because of industrialization or the location of industrial estates nearby, large-scale manufacturing has not yet affected the vast majority of secondary cities. In most, the manufacturing sector remains relatively small and is composed of small-scale establishments, most employing less than 10 workers.

As noted in Chapter 2, there is often a discernible difference in the structure and characteristics of the industrial sector in the primate or capital city of most developing countries and those of even the largest secondary cities. Studies that compare Nigeria's capital city, Lagos, with the secondary metropolis, Kano, for example, indicate that the industrialization of the two cities have been quite different. Investment by multinational corporations has been concentrated primarily in the capital, with far less investment in Kano. Production processes in the secondary city are more labor intensive, management is more personalized, labor unions have less influence, labor is less skilled and educated, and the manufacturing sector is more heavily dominated by small-scale producers.[53] When import substitution and petroleum industries began to play a larger role in the Nigerian economy, their locational preference for port cities greatly influenced the growth and diversification of Lagos. But despite the fact that larger secondary cities such as Port Harcourt, Kaduna, and Kano developed strong competitive advantages over smaller cities in attracting capital-intensive industries, these firms play a less important role in the economies of secondary cities than they do in the capital city.[54]

Most of the secondary cities in developing nations grew without benefit of large-scale industrialization. Muench observes in his study of Ibadan, Nigeria, for example, the "almost complete absence of large-scale industrial and commercial activity for a city of its size. Instead most of the economic vitality of the city stems from thousands of small-scale enterprises," ranging from those traditional operations that provide their owners with small amounts of income to more modern establishments that return a comfortable profit.[55] Weiker's study of Eskisehir, Turkey, noted that the percentage of population employed in manufacturing in that provincial city was similar to the largest metropolitan centers, but that the structure of manufacturing employment was far different. During the 1960s and 1970s, more than 90 percent of the manufacturing plants in Eskisehir were small-scale handicraft establishments; few hired more than 10 employees or used machinery for production.[56]

Most secondary cities with less than a million people in Latin America also have manufacturing sectors dominated by small-scale establishments. Roberts reports that in Huancayo, Peru, only 2 large industries existed in the 1970s—a textile mill and a knitwear factory that together employed about 300 people—and that the average number of employees in the city's other manufacturing establishments was 7. More than half of the labor force in the manufacturing sector was employed in enterprises with less than 5 workers. Small-scale industry is organized and operated in Huancayo much like small-scale commercial activities are in most intermediate cities: They are family owned and run, they are operated in an informal and traditional manner, they are labor intensive, and they produce goods at relatively low cost that can be bought by low-income families.

Even when secondary cities have a few large industrial plants, it is not clear that these plants have a substantial impact on the economy or that the benefits trickle very far down the income scale. Selby and Murphy observe that the rapid industrialization of San Luis Potosi in Mexico primarily benefited those skilled workers who already had jobs, for they, rather than the unemployed, received the better positions in the new factories.[57]

In Asia, even secondary cities in industrializing countries have manufacturing sectors dominated by small-scale establishments. In Chiangmai, Thailand, for instance, about 43 percent of the labor force of the province was employed in manufacturing in the late 1970s; but 90 percent of the industrial establishments were agricultural processing activities employing an average of less than 10 workers. Larger firms employed an average of 17 workers each.[58]

However, the manufacturing sector in secondary cities can play an important role in providing income and employment and establishing a base for industrial deconcentration within the country. The experience in Taiwan indicates that under favorable conditions secondary and small cities can support a substantial number and variety of manufacturing establishments. Ho points out that between 1930 and 1956 industrial employment increased by nearly 4 percent a year in Taiwan's 7 largest cities, and that between 1956 and 1966 it increased by about 5.6 percent a year in secondary cities outside of the Taipei metropolitan area.[59] Employment in food, textiles, furniture and fixtures, nonmetallic mineral products, and machinery and equipment increased from 3.0 to 6.9 percent a year, in chemicals by over 7 percent, and in wood and metal products by more than 12 percent a year in secondary cities. By 1966, Kaohsuing, Taichung, Tainan, and 23 urban townships adjacent to these secondary cities had 23 percent of Taiwan's employment in manufacturing and 8 smaller cities had an additional 10 percent. Ho argues from his review of experience in Taiwan that a decentralized pattern of industrialization based in secondary and smaller cities in rural areas improved nonagricultural employment opportunities for rural households by allowing them to commute to manufacturing jobs in nearby cities and still take part in farm activities on weekends and created numerous linkages with small commercial, service, and repair establishments in rural towns, making it possible for small businesses to develop in rural areas. Decentralized industrialization created employment and entrepreneurial opportunities for rural people, giving them more income to spend on manufactured goods produced in secondary cities.[60]

That secondary cities in most developing nations are primarily small-scale industrial centers does not imply that large-scale industrialization has not influenced the growth of some secondary cities or that they cannot accommodate large manufacturing firms. As in Taiwan, secondary cities have played an important role in deconcentrating some industries from the largest cities in India, Brazil, Nigeria, and South

Korea. In some Nigerian cities physical expansion and population growth were due almost entirely to the location of industrial estates nearby, to which access roads, water, electricity, and other utilities were extended. Industrial estates in Kaduna, Kano, and Port Harcourt became nuclei for urban growth and expansion. [61] In South Korea manufacturing firms in five secondary cities with more than a half million residents employed more than a million people in 1980, almost double the number working in industries in those cities in 1974. More than a half million people were employed in industry in smaller secondary cities. In 1974, Korean secondary cities contributed about 54 percent of the country's value added by manufacturing, and, although no individual city's manufacturing sector added as much as that of Seoul, the secondary cities together contained a substantial amount of Korea's manufacturing capacity. [62]

The Korean experience shows that secondary cities can play an important role in countries where government seeks to deconcentrate manufacturing employment from the largest metropolis. By 1980, although small-scale establishments still accounted for about 90 percent of all industrial firms in secondary cities, industries employing 100 or more people engaged more than 50 percent of the industrial labor force in nearly half of the Korean cities with 100,000 or more residents. In Pusan, Incheon, and Daejeon—3 of the 5 largest secondary cities—large factories employed from 55 to 68 percent of the industrial workers. In smaller secondary cities, where the government has created industrial estates, a large majority of the manufacturing work force is employed by large-scale industries: in Masan, 82 percent; in Ulsan, 80 percent; in Cheongju, 61 percent; in Pohang, 70 percent; and in Chuncheon, a little more than 50 percent. In Iri, more than 61 percent of the manufacturing labor force was employed by large firms, as were 73 percent of the industrial workers in Gunsan, 61 percent in Weonju, and 76 percent in Andong. By 1980, the average number of workers in industrial firms in cities of more than 200,000 population was 85, and the large firms had an average of 54 employees each in cities with between 100,000 and 200,000 residents. In only one-third of Korea's secondary cities did more than 60 percent of the manufacturing labor force still work in small-scale industries in 1980 (see Table 4.4).

Korea's policy of extending highways, providing utilities, upgrading power and energy capacity, and establishing essential infrastructure allowed these cities to support large-scale industry successfully. By 1980, half of the smaller secondary cities had more than 5 large factories, as did all of the cities with populations of between 200,000 and 500,000.

In other countries, the increasing agricultural production is not only the result of growing demand for farm products in secondary cities, but in turn creates new demands for manufactured goods and increases employment opportunities in the urban centers. The increasing agricultural production in the Punjab region of Pakistan during the 1960s that came from the adoption of "Green Revolution" technology generated increased demand for agricultural equipment, machinery, and supplies in seven secondary cities in the Punjab. Increased agricultural production spawned investment in diesel engine and other irrigation equipment production in Lahore and Daska, and in tubewell components, cane crushers, fodder choppers, pipe, and farm implements in Multan, Sahiwal, Gujranwala, Gujrat, and Lyallpur. The vast majority of firms manufacturing agricultural equipment in these secondary cities employed less than 15 workers, but many were linked to larger firms that produced components assembled by smaller establishments and to artisanal and craft shops that provided other agricultural equipment parts and components.[63]

Small-scale industry seems to be both a strength and a weakness of the economies of secondary cities. Small-scale manufacturing does not generate large amounts of employment for those outside of the families that own and operate

TABLE 4.4 Characteristics of and Employment in Manufacturing Establishments in Secondary Cities of South Korea: 1960-1980

City	Population (in thousands) 1978	Number of People Employed in Manufacturing (in thousands)			Number of Manufacturing Establishments		% of Value-Added by Manufacturing in Secondary Cities 1974	% of Firms and Employment by Firm Size			
		1960	1974	1980	1974	1980		Large	Employment	Small	Employment
Pusan	2,879.6	57.7	250.2	665.7	1,274	4,541	10.9	3.1	55.3	96.9	44.7
Taegu	1,487.1	47.5	134.9	120.3	783	2,957	5.2	1.2	32.5	98.8	67.5
Incheon	963.5	16.6	69.2	181.9	481	1,579	6.2	11.7	54.7	88.3	45.3
Gwangju	694.6	10.4	28.9	24.1	114	505	0.8	1.3	41.3	98.6	58.7
Daejeon	580.6	10.9	37.0	72.9	250	926	1.7	50.9	68.3	49.0	31.6
Masan	391.6	5.7	35.7	58.9	191	128	2.7	6.3	82.4	93.7	17.6
Jeonju	384.1	6.8	16.8	13.2	91	170	1.8	4.7	37.5	95.2	62.5
Ulsan	364.5	.8	24.5	74.2	53	144	8.4	22.9	80.9	77.1	19.1
Seongnam	324.1		22.7	47.5	107	450	0.9	1.3	22.4	98.7	77.6
Suweon	266.1	3.7	20.5	34.6	82	256	4.7	3.5	66.7	96.4	33.3
Cheongju	223.1	3.5	10.9	17.4	59	146	1.2	21.2	61.4	78.8	38.6
Mogpo	210.9	4.2	10.3	9.2	60	201	0.4	2.5	52.1	97.5	47.9
Anyang	187.9	2.9	14.9	42.0	104	374	1.9	5.3	49.5	94.7	50.5
Pohang	184.0	1.6	10.9	28.4	34	129	1.9	8.5	78.7	91.5	21.3
Jinju	174.9	3.5	10.5	12.5	72	240	0.3	2.5	34.3	97.5	65.7
Gunsan	167.4	4.2	12.1	35.2	70	125	0.9	6.4	73.1	93.6	26.9
Bucheon	163.5	2.7	9.9	53.2	103	837	1.0	2.2	11.1	97.8	88.9
Chuncheon	152.6	2.1	5.1	5.8	29	93	0.1	4.3	51.2	95.7	48.8
Jeju	152.5	2.1	4.2	3.6	28	52	0.1			100.0	100.0
Yeosu	151.3	1.8	6.0	3.8	46	99	0.1	3.1	23.1	96.9	76.9
Iri	132.3	2.4	8.3	2.3	61	176	0.3	7.9	61.4	92.1	38.6
Weonju	131.0	1.8	6.6	3.8	22	113	0.1	1.8	16.5	98.2	83.5
Eujjeongbu	117.8	1.1	6.3	7.4	22	28	0.3	17.9	46.6	82.1	53.4
Suncheon	114.6	1.5	2.9	1.4	9	98	0.3			100.0	100.0
Gyeongju	113.9	1.5	3.2	3.7	12	58	0.1	1.7	19.9	98.3	80.1
Chungju	110.1	2.3	6.3	5.3	14	88	0.4	3.4	64.3	96.6	35.7
Cheonan	109.3	1.7	6.3	18.7	39	113	0.2	4.4	49.1	95.6	50.9
Jinhae	108.7	1.0	6.9	3.4	11	52	0.5	5.8	56.5	94.2	43.5
Gangneung	102.2	1.4	3.3	1.5	16	14	0.1			100.0	100.0
Andong	101.5	1.4	3.7	1.8	21	97	0.2	23.7	76.6	76.3	23.4

SOURCES: See source note, Table 4.3.

them, and returns only low levels of profit to most entrepreneurs. Small firms usually do not generate large amounts of tax revenue for local governments and they are risky ventures from which total output is usually low. Yet, collectively, they employ relatively large numbers of people, provide low-cost goods for those with low incomes and for other small-scale businesses and industries, offer opportunities for expansion and further investment to the more successful entrepreneurs, allow for deconcentration of industry within a country, and provide a base upon which to diversify and build manufacturing capacity.

Regional Transport and Communications Centers

Many case histories of secondary cities emphasize the importance of transportation and communications to their development and to their contemporary roles as regional marketing and trade centers. In Turkey, for example, Weiker notes the large number of bus companies providing frequent service from Eskisehir to larger metropolitan areas such as Istanbul and Izmir as well as to other intermediate cities such as Bursa. The city is also the main junction for the Istanbul-Ankara and Istanbul-Konya railroads.[64] Tripoli, Lebanon, achieved its status as a regional trading center because of its extensive port facilities, paved highways to Beirut, rail connections to Beirut and various cities in Syria, and bus connections to cities in Syria and Egypt as well as to other cities and towns in Lebanon. Tripoli grew not only as a regional and national port but as an international port, distributing agricultural products grown in the northern part of Lebanon to Europe, Africa, and the Middle East.[65]

As noted earlier, transportation networks also played a crucial role in the growth of towns into secondary cities in much of Africa. Their role as transportation and communications centers is still crucial to the economies of African cities, especially in Nigeria, where Mabogunje observes that the railways and road networks not only created new secondary cities but also transformed the economies of existing ones from administrative and marketing to commercial, service, and industrial centers.[66]

In Latin America, transportation and communications facilities were essential to the growth of marketing centers. Beals notes that the modernization of transportation systems and the extension of highways in and around Oaxaca, Mexico, stimulated changes in urban and rural markets and allowed them to achieve greater efficiency. The highways increased the access of rural producers to the urban marketplace. Because of their ability to use trucks more effectively, traders could carry more goods and visit more villages after the highway system was extended through Oaxaca. The highways not only extended the city's market area further into the rural hinterland, but also increased intervillage trade.[67]

Many of the secondary cities in Asia are transport and communications centers for their regions. In Indonesia, for example, Medan is at the axis of the north-south road and rail lines connecting the city to interior areas of northern Sumatra. Railways extend to Aceh and roads connect Medan to provincial and interprovincial highways. The city has both a harbor and a port, with the harbor serving as the hub of regional exchange and trade, and it became a terminal for intercity bus service in the late 1960s. Moreover, Medan has telephone exchanges, more than 80 public and private radio stations, and a television channel. Ujung Pandang (Makassar), also a port, is well connected by roads to the interior of South Sulawesi; its airport links the city to others in Java and Bali. Like Medan, it is a center of modern communications, with more than 95 radio stations and telephone and telegraph service, not only for its own residents but for those of rural towns and villages within traveling distance of the city. Both Padang and Semarang are port cities, the latter with east-west rail connections and north-south road and highway linkages. Padang is the center of air, sea, rail, and bus service for much of the population of West Sumatra.[68] The role modern highway and transport facilities play in linking Chiangmai to Bangkok and to other towns and villages in northern Thailand and in making the city a major trade, tourist, and service center has already been noted, but the importance of internal transportation systems to the growth of secondary cities also deserves to be mentioned. Commerce, trade, and tourism in Chiangmai are facilitated by an efficient mini-bus system that provides 70 percent of the transportation services within the city.[69] Pannell notes also that an important factor in the growth of Taichung, Taiwan, is that the "transport system is fast, efficient and cheap."[70]

Finally, an indication of the importance of transport, communications, and related activities is reflected in the number of jobs they provide and the contribution they make to the regional economy. In Thailand's two largest secondary cities, the transport sector has been growing rapidly. In Chiangmai, it contributed nearly 6 percent of the share of gross provincial product in 1976, and grew by more than 12.5 percent between 1970 and 1976; in Songkhla, transport contributed nearly 8 percent of the gross provincial product and grew at about the same rate as in Chiangmai—a rate that was three times greater than Bangkok's during the same period.[71] By the mid-1960s three secondary cities and their surrounding townships in Taiwan provided nearly a quarter of the transport and communications employment in that country, and together with eight smaller cities accounted for nearly one-third of Taiwan's employment in the sector.

Employment in transport and communications grew by more than 5 percent a year in Taiwan's secondary urban centers from the mid-1950s to the mid-1960s.[72] In Davao City in the Philippines, transport and communications establishments employed nearly 9 percent of the city's labor force in 1978.[73] And, because many of Indonesia's secondary cities are ports, the transport and communications sector accounted for a significant share of their employment during the 1960s and 1970s.[74] In his study of Meerut City, in India, Sundaram points out that employment in transportation, communications, and storage firms in that urban center increased by 71 percent between 1961 and 1971—the second highest increase in employment for any sector in the city's economy—to account for nearly 10 percent of Meerut's work force. The growth of transportation and communications employment was a reflection of the growth of manufacturing and commerce in the city. The transport and communications sector became the third largest employer during the 1970s.[75]

Although the variations were great among Korean secondary cities, those with 500,000 or more population had an average of nearly 9 percent of their labor force working in transportation, communications, and related activities in 1980; cities with between 200,000 and a half million people had an average of 7 percent, and those with from 100,000 to 200,000 residents had an average of 8 percent of their labor force employed in the sector. The sector was a more significant source of employment for Incheon, Jeonju, Daejeon, Cheongju, Jeju, and Suncheon, where 12 percent or more of the labor force was engaged in those activities, and in Gangneung, a small intermediate city in northeastern Korea, where 35 percent of the labor force was employed in the sector. Nearly 130,000 people were employed by transport, communications, and storage firms in Korea's secondary cities in 1980. The distribution of firms and employment among secondary cities, however, was highly skewed toward the largest; over 60 percent of the firms in this sector, and about 68 percent of the workers, were concentrated in the 5 largest intermediate cities.[76]

*Centers of Attraction for Rural Migrants
and Sources of Income Remittances*

Although the nature and impact of their functions in absorbing rural migrants and providing income remittances to rural areas are less well defined than many other roles played by secondary cities in developing countries, some evidence suggests that these cities can perform potentially important roles in changing urban population distribution and providing income to rural areas.

That secondary cities have been attractive to rural migrants as temporary or permanent homes is obvious from their growth, only about half of which is thought to be due to natural increase. Yet their growth rates and the size of their population increases over the past 25 years indicate that these cities have been less attractive than national capitals and the largest metropolises. Not only do they seem to attract a smaller share of rural migrants in many countries, but the distances from which they draw migrants tend to be shorter than those of the largest cities.

In much of Asia, as noted in Chapter 2, the highest rates of population growth were found in the largest cities during the 1950s and 1960s. Cities of a half million or more people recorded the highest growth rates in Taiwan, Korea, Indonesia, the Philippines, and Pakistan. They grew by more than 10 percent a year in Taiwan, 15 percent in Korea, and nearly 6 percent in Indonesia between 1950 and 1970. Cities of smaller size grew much less rapidly, and in Taiwan, Indonesia, the Philippines, and India, cities smaller than 100,000 lost population.[77] Studies of urbanization and migration in Latin America found similar trends in the growth of the largest cities, with smaller urban centers either not absorbing migrants at a rapid enough rate to slow the growth of the largest metropolises, or losing population to larger secondary and primate cities.[78] Studies of migration in Indonesia indicate that secondary cities have an area of attraction much more restricted than that of the largest metropolitan center. Hugo notes that a large proportion of secondary cities' permanent migrants is drawn from their own provinces or an even more restricted area. This fact is reflected in their ethnic and linguistic characteristics;[79] many secondary cities are dominated by a single ethnic or linguistic group.[80]

As Korea's third largest city, Taegu, was growing rapidly during the 1950s and 1960s, its strongest attraction was for people living within the same province. More than 70 percent of the migrants sampled by Barringer came from immediately surrounding rural areas, and an additional 15 percent came from an adjacent province. Less than 14 percent came from other provinces in Korea, indicating Taegu's restricted field of migration. In Taegu, as in other secondary cities of Korea, migration flows have been dynamic—better-educated, higher-income, and more adventurous people leave the smaller cities after a short period of time for Seoul to pursue better economic opportunities, and younger and more ambitious people from the rural areas make their way to Taegu and other secondary cities to pursue similar opportunities.[81] An exception to this restricted field of migration for secondary cities seems to be in those that grew primarily because of industrialization. Iligan, in the Philippines, for example, "generated large-scale migration of thousands of skilled and unskilled workers and professionals from all corners of the archipelago," according to Ulack.[82]

MIGRATION TO SECONDARY CITIES

In any case, studies of a number of secondary cities indicate that they can and do draw rural people and that they can have an effect on the distribution of urban population by offering locations other than the primate city for settlement. To the extent that secondary cities offer more job opportunities and better educational facilities and social services than rural areas, they can become stopping-off points for migrants who might otherwise go directly to the largest metropolis. For many migrants, the stop off becomes permanent. Evidence from Brazil indicates that secondary cities there have been able to attract and retain substantial numbers of rural migrants. Census data from 1970 suggest that about 67 percent of rural and 68 percent of urban males who migrated to 7 secondary cities—Belo Horizonte, Porto Alegre, Curitiba, Salvador, Recife, Fortaleza, and Belem—in 1965 were still living there 5 years later, although only about half of the male migrants who came to these cities from 1959 to 1965 were still there in 1970. Short-term retention rates for female migrants were slightly lower, but long-term rates were approximately the same as for males. Although the retention rates of secondary cities were lower than those of São Paulo and Rio de Janeiro by more than 10 percent, the intermediate cities did retain about half of all migrants that initially moved to them. Retention rates in the primate cities also decreased over time, but not as drastically as in the secondary metropolises.[83]

Secondary cities seem to offer opportunities for upward mobility and economic improvement for many who remain. Studies of northeastern Brazil conclude that although many of the urban migrants in that poor region of the country remain in poverty, their economic situation often improves and that they make an important contribution to the dynamics of urban development. Geiger and Davidovich argue from the study of migrants to Natal that "in spite of their low income, they contribute more effectively to broadening the urban market than they did in the rural state they previously occupied. Hence to the extent that a city like Natal attracts manpower formerly engaged in agricultural pursuits, it withdraws such labor from a predominantly subsistence regime and inserts it in a monetary economy."[84] These authors conclude that the growth of secondary cities in northeastern Brazil has created some degree of internal equilibrium in the region, absorbing rural migrants there and slowing outmigration to São Paulo and Rio de Janeiro.[85] Studies from Asia and Africa conclude that retention rates are often lower in secondary cities than in larger metropolises because many migrants must find employment in the informal tertiary sector or in small-scale commercial activities, which have low labor-absorption capacity. Studies of migrants to cities with more than 75,000 population in Peninsular Malaysia show that more than 40 percent of those who migrated before 1965 were employed in services, nearly 21 percent in commerce and 11 percent in transportation, communications, and utilities. Nearly 73 percent of those who had come to secondary cities in the 5 years before the 1970 census were working in tertiary activities. However, many of those who remained and supported themselves through informal or service sector employment later moved into commercial, technical, or clerical occupations or into the wage service sector and began to achieve some level of upward mobility in secondary cities.[86] Similarly, Lubeck's studies of Kano, Nigeria, show that most of the migrants initially found work in tertiary activities, but that many of the informal activities were partially integrated with modern commercial and manufacturing establishments, and that this also provided a channel of upward mobility. The informal sector in Kano, and other African secondary cities, provides services to modern establishments at low cost and has been a channel for the recruitment of factory workers. In Kano, most factory workers from rural areas were previously engaged in the informal sector; they loaded trucks, pushed oil drum carts, practiced traditional crafts such as shoe repair, sold used clothing and snacks, repaired appliances and household equipment, and sold a variety of goods.[87]

Many secondary cities share with the largest metropolitan areas the characteristics that make cities attractive to rural people. Those with strong or growing industrial sectors seem not only to attract people from greater distances but also to draw people with particular socioeconomic characteristics. In Taegu in the 1960s and 1970s, and in many other secondary cities in the developing world today, economic factors motivate people to move from rural areas. Over 50 percent of those migrants to Taegu who were surveyed during the 1960s moved to the city to escape poverty in the countryside, to seek employment, or to take advantage of business opportunities in the city. Nearly 10 percent more moved to obtain a better education. Those migrants who were attracted to the city during its period of rapid growth were relatively young; nearly three-fourths of the migrants entering Taegu during the 1960s were under 30 years old, and many were in their late teens or early 20s. Nearly 60 percent came directly from the countryside and a little more than 31 percent came from smaller towns and cities. For them, Taegu presented opportunities they did not have in rural villages and small towns.[88]

Similar characteristics were seen in migrants to Iligan during the 1970s. Ulack points out that most migrants to this Philippine city were young—nearly 60 percent of those he interviewed were between 15 and 29 years old. The median age of male household heads was 23.9 years and that of female migrants was a little over 21 years. Many were better educated or more skilled than the average rural Filippino; and many came, as do migrants to cities in many developing countries, because they already had friends and relatives working in the city.[89]

REMITTANCES TO RURAL AREAS

Secondary cities also seem to play an important role in channeling remittances earned by rural migrants to their home villages, although the magnitude is not known for any developing nation. Studies in Kenya indicate that about 13 percent of the income earned by men in a sample of recent migrants to Kenyan urban centers was remitted home; surveys of India report that as much as 38 percent of the income earned by migrants working in urban factories is returned to their rural villages; and studies of Thailand suggest that nearly half of the migrants surveyed intended to take money with them when they returned to their rural homes.[90] In some rural areas of Nepal more than half of the households receive money from relatives who have migrated to Kathmandu. In Ghana, 40 percent of rural households with members in cities have received cash or in-kind remittances. Studies of 48 rural villages in India found that remittances were received by up to 40 percent of the households.[91]

Rempel and Lobdell conclude from their review of African developing countries that the proportion of income earned by migrants in Third World cities that is returned to rural villages varies greatly among countries, depending on the strength of social and economic ties between migrants and their families, and on such factors as the amount of income the urban dwellers earn and their length of stay in the city. Whether or not the migrants bring their immediate families with them to the cities also affects the flow of remittances. However, even when remittances are large, they tend to become smaller as migrants remain in the city for longer periods of time and as their ties to their home villages become more tenuous.[92] The money sent back from cities seems to be used by rural relatives to meet social obligations, to pay for marriages, school fees, house repairs, support of elderly kin, and investment in productive activities. Moreover, a good deal of evidence indicates that the flow is not simply one way, but that during their initial period of residence in the city, migrants receive money from relatives in rural areas to support them until they find jobs or can earn enough to survive in an urban area.[93] Whatever the magnitude of the flow, it is clear that secondary cities are channels through which money from rural areas comes to migrants seeking urban employment, and earnings from migrants in the city are sent to rural villages and towns to help support family and kin.

Centers of Social Transformation

Secondary cities also seem to play important roles in fostering changes in attitudes and behavior and in easing the transition from rural to urban life for people who migrate from farms and villages. Among their most important social functions are that they encourage and accommodate social heterogeneity and provide an environment in which diverse social, ethnic, religious, and tribal groups can be assimilated into urban society. Secondary cities accommodate a variety of organizations that socialize rural people in urban areas, support them during their transition to urban living, and mediate conflicts among them. The social interaction that takes place in secondary cities infuses new attitudes, behavior, and lifestyles that are more conducive to urban living and to coping with urban problems. Moreover, secondary cities provide opportunities for economic and social mobility and can offer new economic and social opportunities to women.

SOCIAL HETEROGENEITY AND ASSIMILATION

One of the most frequent observations in the case histories of secondary cities reviewed here is that these cities differ discernibly from smaller towns and villages in their social composition, their mix of urban and rural lifestyles, and their ability to assimilate different social, religious, ethnic, and cultural groups. Hazelhurst observes that in India, for example, "the multiplicity of economic functions in the middle range city is complemented by a corresponding lack of social homogeneity among those who perform these functions. Thus, middle range cities differ from other cities where trade and manufacturing tend to be controlled by particular social groups, such as a single merchant class."[94]

The secondary city's role as regional market center is one means by which it mixes rural and urban people, first through economic exchange and then through social interaction. First contact with the city and with urban lifestyles often comes to many rural residents through market trade. Srivastaba observes that in the periodic markets (*hats*) of India's industrial city of Ranchi,

> apart from commercial exchange, and an opportunity to convert kind into cash, the urban *hat* provides the occasion for recreation, e.g., cock-fight, or matrimonial engagement, or visit to the city and seeing objects of modern material civilization. The *hat* has been a medium of cultural contact and impact. In the *hat* the tribal is partly in his own natural environment because of his own ethnic groups transacting business and the produce of his countryside so prominent there, and yet the external world is also present there in the form of the urban visitor and the urban and imported commodities.[95]

The social heterogeneity of, and interaction among, people living in secondary cities comes more directly, perhaps, from the city's role in absorbing rural migrants. Attracted to secondary cities because of the economic opportunities they seem to offer, different groups come into contact with each other and must live together in some degree of harmony simply to survive. Ulack points out that as Iligan grew as a commercial and manufacturing center, its population became "among the least homogeneous in terms of cultural composition." He notes that as people from varied economic, cultural, educational, and social backgrounds are brought together "some of the more valuable traditions and ideas of one group become apparent and may in time eventually become accepted by another group."[96]

Thus different social, religious, and ethnic groups are not simply thrown together in secondary cities, but these urban centers support a variety of social organizations that allow people from different backgrounds to interact with members of their own groups and with those of other backgrounds while assimilating into urban society. In the process of adjusting to urban life, the new residents are exposed to, and often adopt, new ways of thinking, behaving, and doing things that allow them to cope with the problems and complexities of urban living and to compete for the economic and social opportunities that secondary cities offer. The richness of the social organization of such cities is noted in many case studies. In the Middle East, Gulick observes that secondary cities such as Tripoli were large enough to allow both a degree of anonymity, for those who sought it, and social interaction in a new environment, through ethnic associations, social organizations, and neighborhood groups that were spawned in the city. Gulick describes the capacity of these organizations to protect people in particular social, religious, or ethnic groups by allowing them to associate with people of their own backgrounds. The associations are necessary, Gulick points out, because "to each resident of Tripoli the vast majority of Tripolitans are strangers."[97] People can insulate themselves or seek broader associations with others who are similar or different.

For newly arrived migrants from rural areas these organizations offer shelter and opportunities for jobs and contacts; they mediate conflicts and maintain communications among their members and between them and people who have risen to positions of influence in the city who are sympathetic to their advancement. Ethnic associations in African cities, for instance, play a variety of roles: They help migrants without families in the city to find housing, jobs, capital to establish businesses, and loans to help them over difficult times or to take advantage of promising opportunities to increase their incomes. In much of the Middle East and in many secondary cities of Africa, Islamic institutions, for instance, not only serve as channels of acculturation into city life, but also provide subsistence for temporary or permanent migrants, find jobs for rural youth studying in the city, and recruit labor for their members who have businesses. Lubeck observes in his study of Kano, Nigeria, that

> when a student arrives in a city like Kano, he often has a letter of introduction, the name of a *mallam* or a student of their mallam, or in some cases the hospitality of a local Koranic student whom they met in a rural school. Younger student migrants usually study in the morning and evening and work at casual labor during the late morning or afternoon. If there are no employment opportunities, their subsistence is provided by the Islamic obligation of alms-giving. If they possess no lodging, they are allowed to sleep in the *zaure* or entry room which nearly all Hausa compounds possess.[98]

In some ways the social functions of these organizations substitute for the close family and community ties often found in rural villages. Both in Kano and in Tripoli, for example, Islamic, Christian, and other religious organizations introduce migrants to new friends, provide help to the sick, bury those who die, and celebrate at weddings or christenings. They offer opportunities for their members to associate with other groups in the city, impose widely shared rules of behavior, mediate disputes and conflicts, and help repatriate those who fall into destitution. The associations may also provide a channel for remitting income or mobilizing capital for investments in members' home villages or towns, and, by maintaining contacts with the areas from which members came, disseminate new ideas and innovations in rural areas.[99]

Urban living also creates pressures that bring changes in family structure: a shift from extended to nuclear families, a weakening of traditional family obligations and rituals, and a tendency toward having fewer rather than more children. In larger secondary cities, such as Tripoli, these changes were already apparent in the early 1960s. Gulick noted that nuclear families were dominant in Tripoli by then, and that

even extended households tended to have only a few relatives, rather than three or four generations, living together.
Some traditional obligations were maintained and others
were discarded or infrequently practiced.[100]

In smaller cities, or in those in the early stages of urbanization, and for many of the poorer residents and recent
migrants, however, the extended household is one source of
sustenance and survival. Studies of Taegu, Korea, in the
1960s note that "the extended family was an especially
important social institution," for it provided most of those
who came to the growing metropolis with shelter, basic
needs, and protection.[101] Among the poor and for some
middle-income groups, there are strong pressures against
maintaining extended families, but some advantages to having
extended households. Selby and Murphy, in their studies of
Mexican intermediate cities, and particularly of Oaxaca,
point out that "the householder who manages to keep his
family together as a contributing cooperative will do far
better than the householder who permits his family to split
into its constituent nuclear groups."[102] Even among the
poor, however, the lifestyle in secondary cities makes patri-
lineal extended families difficult to maintain over long
periods of time, and competing pressures are felt among the
poor.

There are even stronger pressures on middle-income families in many secondary cities to limit the number of children.
Studies of intermediate cities in Mexico conclude that the
direct contribution to well-being from additional children is
negative, for dependent children add to household expenses
in the city without providing the labor they often contribute
in rural areas. There are incentives for middle-income families
in secondary cities to maintain extended households, however, if this adds to the number of income earners. Even for
middle-income families in cities with a low-wage employment
structure, the extended household containing more adult
members tends to produce a higher total income and a

relatively better standard of living. Indeed, a distinguishing characteristic of middle-income households in many secondary cities of Mexico is their greater ability, compared to the poor, to place a second member of the family in the work force. In Oaxaca, 37 percent of middle-income families had a second worker, compared to only 19 percent of the poor. In San Luis Potosi, nearly half of the middle-income households have more than one member employed, compared to 9 percent of the poor; in Mexicali, three times as many middle-income families had more than one worker than poor families.[103] Thus the pace and extent of change in family structure, and the implications for family size and labor force participation, depend on income distribution as well as on economic structure in secondary cities of developing countries.

PROMOTION OF SOCIAL CHANGE AND CREATION OF NEW ROLES AND OPPORTUNITIES

With growth and diversification come a number of changes in the societies of secondary cities that differentiate them from small towns and rural villages. The variety and pervasiveness of social changes that occurred in Chiangmai, Thailand, not all of which were considered desirable by long-time residents, are described by Noranitipadungkarn and Hagensick. Traditional northern cultural patterns, behavior, dress, and beliefs were slowly diluted, the pace of life quickened and people became more competitive, local neighborhood interaction weakened, participation in temple activities among younger people declined, youth sought activities and recreation outside of the family circle, and distrust of strangers became more widespread as modernization occurred in Chiangmai.[104]

Debates over the value and desirability of these changes make up a large part of the commentary in local newspapers and contribute to the tensions that seem to arise between older and younger generations. Often, however, secondary cities are large and diversified enough to accommodate both tradition and change. Many case histories describe the adaptations in social arrangements, land uses, organizational structures, and patterns of behavior found in secondary cities that allow people to cope with urban living. What to the Western eye may seem to be a chaotic mixture of urban land uses in many developing countries, compared to the more "orderly" segregation and separation of land uses in many North American and European cities, for example, reflect in part this ability of urbanites in Asia, Africa, and Latin America to adjust to rapidly changing social and economic conditions. The mixture of land uses in most Asian cities, for instance, serves important social functions. Pannell points out in his study of Taichung, Taiwan, that mixed land uses allow people to watch over and protect their possessions, and prevent the incursion of crime within a neighborhood:

Why does a jeweler on the main street of Taichung prefer to live above his shop rather than in a single-family dwelling one-half mile or more away? He can watch over his shop at night, he can ask his wife or son to attend to it if he feels ill or tired; he can keep his shop open late at night and open and close it very conveniently; he can take his meals easily and comfortably, while he keeps his shop open; he has the advantage of an excellent location for shopping, recreation and attending to his other personal affairs if he so desires.[105]

In cities where transportation is often expensive or inconvenient, mixed land uses can reduce travel time between home and work, and place commercial, service, recreational, and other activities within easy walking or cycling distance to much of the population.

New opportunities that are unavailable or unacceptable in smaller towns and rural villages are also created in secondary cities. In many African and Asian cities, for example, women have new or expanded opportunities to engage in business, trade, or professional activities. Indeed, women have come to play an important role in the commercial life of secondary cities throughout Asia and Africa. Trager points out in her study of Ilesha, Nigeria, that Yoruba women participate actively in local commerce, not only as traders, but also as brokers, intermediaries, wholesalers, and lenders; indeed, "they make decisions about and carry out most of the distribution functions within the internal marketing system."[106] Women's earnings make a significant contribution to family income; those earnings that are not used to buy more goods for trading are put into family savings, used to buy household or personal goods, or used for children's educations.

In brief, the case histories and studies of secondary cities in developing countries suggest that these cities can perform a wide variety of social, economic, and service functions that are important to regional and national development, although all secondary cities do not perform all of these functions and some do not perform them well. The degree to which secondary cities can be important centers in regional and national development depends in part on the degree to which the national government invests in these places to strengthen their economies, and on the commitment of local leaders to making them catalysts for the development of their regions.

NOTES

1. Walter F. Weiker, *Decentralizing Government in Modernizing Nations: Growth Center Potential of Turkish Provincial Cities* (Beverly Hills, CA: Sage, 1972).

2. John Gulick, *Tripoli: A Modern Arab City* (Cambridge, MA: Harvard University Press, 1967), pp. 61-62.

3. Chakrit Noranitipadungkarn and A. Clarke Hagensick, *Modernizing Chiengmai: A Study of Community Elites in Urban Development* (Bangkok: National Institute of Development Administration, 1973), p. 23.

4. Ibid., p. 24.

5. Ibid.

6. Abukasan Atmodirono and James Osborn, *Services and Development in Five Indonesian Middle Cities* (Bandung: Institute of Technology, Center for Regional and Urban Studies, 1974).

7. Richard Ulack, "The Impact of Industrialization upon the Population Characteristics of a Medium Sized City in the Developing World," *Journal of Developing Areas* 9 (January 1975): 203-220.

8. Si-Joong Yu, "Educational Institutions," in *A City in Transition: Urbanization in Taegu, Korea,* ed. Man-Gap Lee and Herbert Barringer (Seoul: Hollym, 1971), p. 450.

9. See Republic of Korea, [Long Range Planning for Urban Growth to the Year 2000: Data Collection], vols. 1, 2 (unofficial trans.) (Seoul: Ministry of Construction, 1980), for comparisons.

10. See Johannes F. Linn, *The Distributive Aspects of Local Government Finances in Colombia: A Review of the Evidence,* World Bank Staff Working Paper 235 (Washington, DC: World Bank, 1976), pp. 5-10.

11. Ibid., p. 46.

12. Atmodirono and Osborn, *Services and Development,* pp. 32-33.

13. For later figures, see Republic of Korea, *Korea Municipal Yearbook, 1980* (Seoul: Ministry of Home Affairs, 1980).

14. John Walton, "Guadalajara: Creating the Divided City," in *Metropolitan Latin America: The Challenge and the Response,* ed. W. A. Cornelius and R. Kemper (Beverly Hills, CA: Sage, 1978), pp. 25-50.

15. See Fernando Greene, "Analysis of Major Mexican Urban Centers 1960-1970" (Ph.D. diss., Cornell University, 1978), especially Table A.8.

16. Harley L. Browning, "Some Problematics of the Tertiarization Process in Latin America," in *Urbanization in the Americas from Its Beginnings to the Present,* ed. R. S. Shaedel et al. (The Hague: Mouton, 1978), pp. 153-172.

17. Greene, "Analysis of Mexican Urban Centers," Tables A.7, A.8.

18. The proposition is tested in a region of Mexico with a city smaller than intermediate size as defined in this study. See B. Lentnek, M. Charnews, and J. V. Cotter, "Commercial Factors in the Development of Regional Urban Systems: A Mexican Case Study," *Economic Geography* 54 (October 1978): 291-308.

19. P. P. Geiger and F. R. Davidovich "Urban Growth as a Factor of Regional Balance-Imbalance," in *Proceedings of the Commission on Regional Aspects of Development of the International Geographical Union,* vol. 1, ed. R. S. Thomas (Ontario: International Geographical Union, 1974), pp. 153-171.

20. Gulick, *Tripoli: A Modern Arab City,* pp. 99-100.

21. V. F. Costello, *Kashan: A City and Region of Iran* (London: Bowker, 1976).

22. Atmodirono and Osborn, *Services and Development.*

23. D. Mookherjee and R. L. Morrill, *Urbanization in a Developing Economy: Indian Perspectives and Patterns* (Beverly Hills, CA: Sage, 1973), p. 59.

24. J. Weinstein and V. K. Pillai, "Ahmedabad: An Ecological Perspective," *Third World Planning Review* 1, no. 2 (1979): 205-233.

25. K. V. Sundaram, *Role of Cities in Attaining a Desirable Population Distribution in the Context of Rapid Urbanization: Case Study of Meerut, India,* Research Project 501 (Nagoya, Japan: United Nations Centre for Regional Development, 1975), p. 104.

26. See Hy-Sang Lee, "An Economic Survey: Efficiency, Equity and Growth," in *A City in Transition: Urbanization in Taegu, Korea,* ed. Man-Gap Lee and Herbert Barringer (Seoul: Hollym, 1971), pp. 187-210; Man-Gap Lee, "Social Organization," ibid., pp. 335-381.

27. Ray Bromley, "Organization, Regulation and Exploitation in the So-Called 'Urban Informal Sector': The Street Traders of Cali, Colombia," *World Development* 6, nos. 9, 10 (1978): 1161.

28. Ibid., pp. 1162-1163.

29. Ibid., p. 1165.

30. See Margaret Peil, "West African Urban Craftsmen," *Journal of Developing Areas* 14 (1979): 3-23.

31. S. V. Sethuraman, "The Urban Informal Sector in Africa," *International Labour Review* 116 (November/December 1977): 343-352.

32. Noranitipadungkarn and Hagensick, *Modernizing Chiengmai,* pp. 19-20.

33. Leighton W. Hazelhurst, "The Middle Range City in India," *Asian Survey* 8, no. 7 (1968): 539-552.

34. Ibid., p. 541.

35. Ibid., p. 549.

36. Norbert Dannhaeuser, "Distribution and the Structure of Retail Trade in a Philippine Commercial Town Setting," *Economic Development and Cultural Change* 25 (April 1977): 471-503.

37. Ibid., pp. 477-498.

38. Ralph L. Beals, *The Peasant Marketing System in Oaxaca, Mexico* (Berkeley: University of California Press, 1975), pp. 120-121.

39. For a detailed inventory and description of these activities, see ibid., Appendix I.

40. Dannhaeuser, "Distribution and Structure of Retail Trade," p. 500.

41. Ibid.

42. Some evidence of this pattern in Africa is found by P. O. Sada, M. L. McNulty, and A. I. Adelemo, "Periodic Markets in a Metropolitan Environment: The Example of Lagos, Nigeria," in *Market Place Trade,* ed. R.H.T. Smith (Vancouver: Centre for Transportation Studies, University of British Columbia, 1978), pp. 155-166; and in Latin America by R. J. Bromley, "Traditional and Modern Change in the Growth of Systems of Market Centres in Highland Ecuador," ibid., pp. 31-47, and Richard Symanski, "Periodic Markets in Southern Colombia," ibid., pp. 171-185.

43. Gulick, *Tripoli: A Modern Arab City,* pp. 97-98.

44. Norantipadungkarn and Hagensick, *Modernizing Chiengmai.*

45. Frederick Temple et al., *The Development of Regional Cities in Thailand* (Washington, DC: World Bank, 1980), pp. 11-12, 76-77.

46. K. V. Sundaram, *Urban and Regional Planning in India* (New Delhi: Vikas, 1977), p. 207.

47. Ibid., pp. 208-209.

48. M. J. Mortimore, "Population Densities and Rural Economies in the Kano Close Settled Zone, Nigeria," in *Geography and the Crowding World*, ed. W. Zelinsky, L. A. Kosinski, and R. M. Prothero (New York: Oxford University Press, 1970), p. 387.

49. Emmy B. Simmons, "The Small-Scale Rural Food Processing Industry in Northern Nigeria," *Stanford University Food Research Institute Studies* 14, no. 2 (1975): 147-161.

50. Sundaram, *Role of Cities*, p. 30.

51. Jin-Hwan Park, "The Growth of Taegu and Its Effects on Regional Agricultural Development," in *A City in Transition: Urbanization in Taegu, Korea*, ed. Man-Gap Lee and Herbert Barringer (Seoul: Hollym, 1971), p. 152.

52. For background information on Korea's agricultural policies, see S. M. Ban, P. Y. Moon, and D. W. Perkins, *Studies in the Modernization of the Republic of Korea, 1945-1975: Rural Development* (Cambridge, MA: Harvard University Press, 1980).

53. Paul M. Lubeck, "Contrasts and Continuity in a Dependent City: Kano, Nigeria," in *Third World Urbanization*, ed. Janet Abu-Lughod and Richard Hay, Jr. (Chicago: Maaroufa, 1977), pp. 281-289.

54. See Akin L. Mabogunje, "The Urban Situation in Nigeria," in *Patterns of Urbanization: Comparative Country Studies*, ed. S. Goldstein and D. F. Sly (Liege, Belgium: International Union for Statistical Study of Population, 1977), pp. 569-641.

55. Luis D. Muench, "Town Planning and the Social System," in *African Urban Development*, ed. M. Kall (Dusseldorf: Bertelsmann Universitätsverlag, 1972), p. 36.

56. Weiker, *Decentralizing Government*, pp. 32-34.

57. Arthur D. Murphy and Henry A. Selby, "Poverty and the Domestic Life Cycle in an Intermediate City of Mexico" (Paper prepared for the Workshop on Intermediate Cities, East-West Center Population Institute, Honolulu, 1980), p. 60.

58. Cited in Temple et al., *Development of Regional Cities*, pp. 75-80.

59. Samuel P.S. Ho, "Decentralized Industrialization and Rural Development: Evidence from Taiwan," *Economic Development and Cultural Change* 28 (October 1979): 77-96.

60. Ibid., pp. 90-92.

61. Mabogunje, "Urban Situation in Nigeria."

62. See Republic of Korea, [Long Range Planning], especially Table IV-4.

63. Frank C. Child and Hiromitsu Kaneda, "Links to the Green Revolution: A Study of Small-Scale, Agriculturally Related Industry in the Pakistan Punjab," *Economic Development and Cultural Change* 23 (January 1975): 249-275.

64. Weiker, *Decentralizing Government*, pp. 32-33.

65. Gulick, *Tripoli: A Modern Arab City*.

66. Mabogunje, "Urban Situation in Nigeria," pp, 583-584.

67. Beals, *Peasant Marketing System*, pp. 222-223.

68. Atmodirono and Osborn, *Services and Development*.

69. Noranitipadungkarn and Hagensick, *Modernizing Chiengmai,* pp. 16-17.

70. Clifton W. Pannell, *T'ai-Chung, T'ai-Wan: Structure and Function,* Research Paper 144 (Chicago: Department of Geography, University of Chicago, 1973), p. 109.

71. World Bank, *Thailand: Urban Sector Review* (Washington, DC: Author, 1978), pp. 79-80.

72. Ho, "Decentralized Industrialization," pp. 80-82.

73. See Robert A. Hackenberg, "A Retrospective View of the Poverty Explosion: Some Results of the 1979 Davao City Survey" (Paper prepared for the Workshop on Intermediate Cities, East-West Center Population Institute, Honolulu, 1980), especially Table III.

74. Atmodirono and Osborn, *Services and Development.*

75. Sundaram, *Role of Cities,* pp. 22-23.

76. See Republic of Korea, *Korea Municipal Yearbook,* Table 36.

77. See Dennis A. Rondinelli, "Balanced Urbanization, Spatial Integration and Economic Development in Asia: Implications for Policy and Planning," *Urbanism Past and Present* 9 (Winter 1979-1980): 13-29.

78. Robert W. Fox, *Urban Population Growth Trends in Latin America* (Washington, DC: Interamerican Development Bank, 1975).

79. Graeme Hugo, "Patterns of Population Movement in Intermediate Cities in Indonesia: An Overview of Some Issues and Examples" (Paper prepared for the Workshop on Intermediate Cities, East-West Center Population Institute, Honolulu, 1980), pp. 19-20.

80. Ibid.

81. Herbert Barringer, "Migration and Social Structure," in *A City in Transition: Urbanization in Taegu, Korea,* ed. Man-Gap Lee and Herbert Barringer (Seoul: Hollym, 1971), pp. 287-334; see especially pp. 293-297.

82.. Ulack, "Impact of Industrialization," pp. 207-208.

83. George Martine, "Adaptation of Migrants or Survival of the Fittest? A Brazilian Case," *Journal of Developing Areas* 14 (October 1979): 23-41.

84. Geiger and Davidovich, "Urban Growth as a Factor," pp. 162-163.

85. Ibid., p. 169.

86. A. H. Hawley, D. Fernandez, and H. Singh, "Migration and Employment in Peninsular Malaysia, 1970," *Economic Development and Cultural Change* 27 (April 1979): 491-504.

87. Lubeck, "Contrasts and Continuity," p. 288.

88. See Yunshik Chang, "Population Growth and Labor Force Participation," in *A City in Transition: Urbanization in Taegu, Korea,* ed. Man-Gap Lee and Herbert Barringer (Seoul: Hollym, 1971), pp. 41-86.

89. Ulack, "Impact of Industrialization," pp. 207-208.

90. A review of experience can be found in Henry Rempel and Richard A. Lobdell, "The Role of Remittances in Rural Development," *Journal of Development Studies 14, no. 3 (1978): 324-341.*

91. J. Connell, B. Dasgupta, R. Laishly, and M. Lipton, *Migration from Rural Areas: The Evidence from Village Studies,* IDS Discussion Paper 39 (Brighton: Institute of Development Studies, University of Sussex, 1974).

92. Rempel and Lobdell, "Role of Remittances," pp. 334-436.

93. See S. M. Essang and A. F. Mabawonku, *Determinants and Impact of Rural-Urban Migration: A Case Study of Selected Communities in Western Nigeria,* African Rural Employment Paper 10 (East Lansing: Department of Agricultural Economics, Michigan State University, 1974).

94. Hazelhurst, "Middle Range City in India," p. 540.

95. A. A. Srivastaba, "Growth, Morphology and Ethnic Character in Ranchi-Dhurwa Urban Complex," in *Urbanization in Developing Countries,* ed. S. M. Alam and V. V. Pokshishevsky (Hyderabad: Osmania University Press, 1976), p. 539.

96. Ulack, "Impact of Industrialization," p. 219.

97. Gulick, *Tripoli: A Modern Arab City,* p. 119.

98. Lubeck, "Contrasts and Continuity," p. 287.

99. Mabogunje, "Urban Situation in Nigeria."

100. Gulick, *Tripoli: A Modern Arab City,* pp. 129-131.

101. Charles G. Chakarian, "Urbanization and Health," in *A City in Transition: Urbanization in Taegu, Korea,* ed. Man-Gap Lee and Herbert Barringer (Seoul: Hollym, 1971), p. 499.

102. Murphy and Selby, "Poverty and the Domestic Life Cycle," p. 23.

103. Ibid., pp. 46-50.

104. Noranitipadungkarn and Hagensick, *Modernizing Chiengmai,* pp. 20-21.

105. Pannell, *T'ai-Chung, T'ai-Wan,* pp. 109, 112.

106. Lillian Trager, "Market Women in the Urban Economy: The Role of Yoruba Intermediaries in a Medium Sized City," *African Urban Notes* 2, no. 3 (1976-1977): 1-9.

CHAPTER 5

STRATEGIES FOR SECONDARY
CITY DEVELOPMENT

The general picture that emerges of secondary cities in developing countries is that they are growing rapidly both in numbers and in population in many nations; that their economies are becoming increasingly diversified, especially in nonagricultural sectors; that they support large numbers of small-scale commercial and manufacturing activities; that they perform a mix of urban and rural functions; and that they provide services, facilities, employment opportunities, and marketing and trade services that are more accessible to a larger number of people than those in smaller towns and rural areas.

Yet the impression also emerges that many secondary cities play a relatively weak role in national development along all of the social and economic dimensions discussed earlier. Large numbers of these cities are concentrated in a relatively few developing nations in all geographical regions. In most countries they account for a smaller percentage of the urban population than the largest metropolitan center, they have grown slowly compared to the largest metropolis, they have absorbed a smaller portion of urban migrants, and their prospects for relieving population pressures on the largest cities seem limited in much of the developing world if current demographic trends hold for the rest of this century. Secondary cities provide substantial amounts of employment in agricultural processing, commercial and service activities, and cottage and artisan industry for their own residents, but have limited capacity to absorb large numbers of rural migrants. Their share of manufacturing is often far smaller than that of the largest cities, and their ability to compete with the major metropolises remains weak. Many remain partially rural in economic base, lifestyle, and physical characteristics; the quality of their social services and facilities is still far below that of the major metropolitan areas; and their share of commercial and service activities often seems to be disproportionately low compared to their share of population.

The Role of National Investment

Why have secondary cities remained a relatively weak component of the urban settlement system in developing nations? The reasons differ from country to country, but one frequent and common reason is that these cities have received a disproportionately low share of national investments in infrastructure, services, industry, and other activities compared to the largest city in nearly every developing nation. The competitive advantage that the metropolises, and especially the national capitals, maintain over smaller cities is not due entirely to superior location or efficiency, but to the fact that national governments in many developing countries have favored the largest cities in the allocation of investments that support industry, create job opportunities, provide better health, educational, and social services, and offer amenities that attract higher-skilled and better-educated workers, foreign and domestic investment, and entrepreneurs.

A study of urban development in Colombia during the 1960s and 1970s, for instance, pointed out "the serious discrimination between the major cities and intermediate cities" in that country in the allocation of both public and private investment. Only 15 percent of total bank and investment corporation financing for manufacturing in Colombia reached the 26 largest secondary cities, although their value-added in manufacturing was about 24 percent and they had 28 percent of the population.[1] Bogotá had a clear advantage over other cities in Colombia in the distribution of tax revenues, import licenses, and administrative determinations that can generate economic growth. The USAID Mission in Colombia noted that the same pattern of discrimination in favor of the largest metropolitan centers and against secondary cities held for financing of infrastructure. Nearly 90 percent of public and private credit went to Bogotá and the 2 largest secondary metropolises, Medellín and Cali. Yet more than half of the total urban deficit in water and sewerage services was in smaller cities. In 1970, the 4 largest departmental capitals received 90 percent of the total municipal revenue budgeted for these cities. The average per capita expenditure for municipal services, education, public health, and economic development assistance was 5 times higher in the 3 largest cities of Colombia than in other cities and generally decreased with city size.[2]

The high levels of population concentration in Mexico City and its commanding lead over other urban centers in manufacturing, commercial, and service activities also did not come about accidentally. They were the result of deliberate policies by successive governments to invest in the capital city as a way of promoting rapid industrialization. The national government, throughout most of the twentieth century, ignored the spatial implications of its economic investment strategy, thereby generating large-scale population agglomeration and industrial concentration in a single metropolitan area. Unikel points out that the government's infrastructure investment policies, heavily favoring Mexico City, "have largely contributed to the present over-concentration of socio-economic activities and population in the MCMA [Mexico City Metropolitan Area] by giving preferential treatment to the metropolis not only in the construction or extension of systems of communications and transportation, power and water supplies, but also in such facilities as hospitals, schools and institutions of higher learning."[3] Once Mexico City reached a position of dominance in the settlement system and economy, even policies aimed at promoting regional industrialization could not counterbalance the enormous attraction of the capital for private investment and, ironically, Mexico City benefited most from incentives designed to foster industrial decentralization. Establishments in the Mexico City metropolitan area captured up to 68 percent of the credit funds granted by the Guarantee and Development Fund for Small and Medium Industry in Mexico in the years between 1953 and 1970.[4]

Similarly, it was not entirely due to its natural advantages that Manila grew to its present dominant size in the settlement system of the Philippines or that it captured the majority of industrial and other productive activities. Nor, given the pattern of government investment among regions of the country, should it be surprising that few other cities in the Philippines have grown as rapidly as Manila and that they are unable to compete with the national capital for commercial and industrial activities. "The unequal distribution of benefits among the population and the unbalanced pattern of sectoral development that characterized Philippine growth for much of the past three decades," World Bank analysts insist, "was closely linked to resource management policies and to the pattern of resource allocation."[5]

The rapid and extensive growth of Metropolitan Manila and its high concentration of productive activities is due largely to its disproportionately large share of national investment. Two regions encompassing and surrounding Metropolitan Manila—Central Luzon and Southern Tagalog—have about one-third of the country's population, yet Javier's studies of regional allocations of government investment show that these two regions have consistently received more than half of government expenditures in all categories of physical infrastructure except highways.[6] In the 1959-1961 fiscal years, for instance, nearly 57 percent of the national government's investments in infrastructure were made in the Metropolitan Manila region, with Southern Tagalog alone receiving more than 49 percent. More than 70 percent of expenditures on ports and harbors, 49 percent on waterworks, 61 percent on flood control and drainage, and almost 70 percent on buildings, schools, and hospitals went to Southern Tagalog. Although the concentration in Southern Tagalog was lessened by the 1971-1973 fiscal years, the allocations to Central Luzon increased substantially, maintaining the Metropolitan Manila area as the most favored. From 1971 to 1973 the two regions split almost evenly 56 percent of all infrastructure investments: 64 percent of port projects, 91 percent of waterworks, 63 percent of irrigation, 67 percent of flood control and drainage expenditures, and 60 percent of building, school, and hospital investments. More than a quarter of all highway outlays were made in these two areas. In the 1959-1961 fiscal years, per capita expenditure on infrastructure in Southern Tagalog was nearly triple that of the rest of the country, and in 1971-1973 it remained at nearly twice that of the average for the country.

Moreover, Central Luzon and Southern Tagalog were favored with higher allocations for social services and government economic development programs. Nearly two-fifths of all community development projects were concentrated in these two regions between 1956 and 1973, as were 43 percent of the allocations for small and medium-size industry and the Board of Investment's large-scale industrial assistance.[7]

This pattern of investment favoring national capitals and large metropolises over secondary cities and smaller towns also characterizes countries in Africa, where urbanization is still incipient. The heavy concentration of public expenditures in capital cities in the Sahel region of Africa has accentuated differences in income and wealth between urban and rural populations generally, and has made it difficult for smaller towns to grow. Studies of expenditure patterns in francophone West Africa show that the high concentration of budget expenditures in capital cities intensified migration from rural areas to the largest city. In a study conducted for the World Bank, Cohen reports that in the Sahel, the pattern of public investment is often an obstacle to the growth of smaller towns. "When compared to the distribution of urban population, despite the importance of the capital cities, it is clear that secondary towns do not receive an equitable share on a per capita basis." The study found that the investment bias toward capital cities applies not only to housing and infrastructure, but also to industry and services. Cohen concludes that the "bias toward the capital cities reinforces economic and spatial patterns which can only be changed through major policy shifts at the national level."[8]

Developmental Versus Exploitational Secondary Cities

If secondary cities are to be strengthened, national governments must be able to distinguish between those that are developmental and those that are exploitational in their functions. Case histories of secondary cities rarely analyze their developmental impact on their population or on their rural hinterlands directly, but a number of the studies provide insights into the kinds of factors that distinguish cities that have promoted widespread development from those that have exploited the resources of their regions. The extent to which secondary cities have "developmental" impacts on their regions seems to depend on the following factors:

(1) the degree to which local elites and leaders identify their own success and status with the economic growth and social progress of the city and its region;

(2) the degree to which local leaders in both the public and private sectors are willing to invest their resources in the growth and development of the city rather than investing surpluses generated from city activities in other places;

(3) the degree to which local leaders and entrepreneurs are innovative and aggressive in introducing more effective methods and techniques of production to increase output and income within the local economy;

(4) the degree to which local leaders and entrepreneurs in both the public and private sectors are aggressive, and successful, in bringing external resources into the city for development;

(5) the degree to which the national government supports the internal growth and development of the city and its region, rather than draining resources from them to support the development of the national capital or the national economy;

(6) the degree to which economic activities established within the city are linked through mutually beneficial processes of exchange to the city's hinterland, thereby serving the needs of rural people and promoting higher productivity and greater distribution of income for the rural population;

(7) the degree to which economic activities are linked to each other within the city to generate "multiplier effects" in investment, employment, and opportunities for entrepreneurship in both large and small enterprises;

(8) the degree to which economic activities within the city are organized to generate income for local residents and promote internal demand for goods and services that can be produced and distributed locally;

(9) the degree to which public and private sectors cooperate in promoting economic activities that generate widespread participation and distribution of benefits; and

(10) the degree to which the city's leaders are willing to promote and encourage—and residents are willing to accept and advance—social and behavioral changes that are responsive to new conditions and needs as they arise.

The effects of these influences can be seen more clearly in a few examples: a city that seemed to have generated economic and social benefits for its own population and for that of its rural hinterland—Chiangmai, Thailand; one that seemed to have exploited its own resources and that of its region during its growth—Huancayo, Peru; and one that has been exploitational, but created conditions for potential development through "anticipatory urbanization"—Davao City in the Philippines.

DEVELOPMENTAL URBANIZATION: CHIANGMAI, THAILAND

In their historical analysis of Chiangmai, Noranitipadungkarn and Hagensick emphasize the catalytic effects that the growth of the city had on the northern region of Thailand during the 1960s and 1970s, and attribute much of Chiangmai's developmental impact to the factors described above.[9] First, there had been, since the early 1930s, a strong sense of identification among local leaders, the social elite, and public officials in Chiangmai with the growth and development of the city. Most of the local public officials, and many from national ministries who were assigned to Chiangmai, regarded their positions as rewarding and important. Assignment to or service in Chiangmai was not seen as an inferior or unimportant post to be held only until one could be appointed to a job in the national capital. Posts in Chiangmai were sought after by senior officials who tended to stay permanently or for long periods of time. Local officials, too, often held office for many years. This created both stability in civil administration and a perception on the part of local and national officials that their success would be measured by the benefits they brought to the city and its region. Moreover, the relatively long tenure of municipal officials allowed them to initiate and see through to completion important development projects.

A majority of the city's entrepreneurs, business leaders, investors, and shopkeepers were born and raised in Chiangmai, and their success was clearly related to the prosperity and continued growth of the city and its surrounding region.

Many of those who were not born in Chiangmai were long-time residents who participated actively in its social, cultural, political, and religious life, as well as in its economic activities. They encouraged and fostered the traditional cultural activities that made life in Chiangmai colorful and that attracted visitors and tourists.

Thus, rather than viewing Chiangmai as a place to make their fortunes to invest elsewhere, business and government leaders directed much of their energy toward bringing in new resources—money, ideas, and skilled people—to promote the city's growth and modernization. During the city's periods of rapid development, business leaders "sought modern ideas through extensive travels that gave them access to relevant information," Noranitipadungkarn and Hagensick point out. "Such travels introduced them to innovative ideas in western style department stores, cultural centers, hotels and improvements in the production of handicrafts."[10] They actively sought new ways of doing things in their Chiangmai businesses. As a result, they invested heavily in mining and tobacco and tea plantations, established a development bank, created credit facilities, built bowling and recreational centers, and revamped cottage industries.

Municipal officials and business leaders were also aggressive in seeking resources from the national government that would improve the city and link it more closely to Bangkok and to other towns and cities in the northern region. They were successful in obtaining millions of dollars for the city during the 1960s and 1970s to build major highways, universities and colleges, hospitals and clinics, electric and water lines, and other public utilities. Local government officials oversaw the building of a sports stadium, a zoo, a large meeting hall, two recreational areas, a modern police headquarters, and a municipal market that served not only Chiangmai's residents but people from neighboring towns and rural areas and that added to the city's attractiveness as a regional and national tourist center. They raised revenues from a variety of sources to build roads and streets that linked the city with arterial highways near Chiangmai and sought funding for new schools, health centers, and libraries.

This developmental spirit among leaders in the city was contagious, and much of the population became caught up in local "boosterism." Both government officials and business leaders promoted the city as a tourist center and invested in activities that would consolidate the city's function as a recreational area. In the 1960s and early 1970s, they built six new movie theaters, a large private medical center, a hospital complex, six modern first-class hotels, six new office buildings, and as many shopping centers.

The investments made in other sectors of Chiangmai's economy were growth generating. They were organized to integrate smaller businesses into the activities of larger establishments and to link rural suppliers with urban producers. Noranitipandungkarn and Hagensick estimate, for instance, that 60 to 80 percent of the people living in rural areas around Chiangmai benefited in some way from the tobacco industries located in the city. The Thailand Tobacco Monopoly opened branch offices in smaller towns and created and supported an extensive network of farmer-suppliers. Thappavong Tobacco Company, the largest producer in Thailand, provided income to 30,000 to 40,000 farmers in the Chiangmai region, supporting them with technical assistance, credit, fertilizer, quality seedlings, and other inputs that helped raise their productivity and income. Tea companies in Chiangmai developed similar relationships with small suppliers in the region and worked with them to improve the quality of their crops. Mining companies provided part-time employment to farmers and urban residents as well as to full-time mine workers. Owners of the timber industries in Chiangmai reinvested much of their profits in the city and sought equipment and services from local suppliers.

Government leaders, too, were sensitive to the relationships between city and countryside. The governor of the province, for instance, spent a great deal of time on his "reaching-the-people" program. Observers note that the governor went to great effort to determine the needs of rural people and to provide government assistance, especially during periods of drought, flood, or cold weather. He used the visits of high-ranking officials from ministries and agencies in Bangkok to obtain food, assistance funds, medicine, seeds, and other resources for distribution to the rural poor in areas around Chiangmai.[11]

There was also a strong sense of cooperation between government and business leaders in the city. Business clearly profited from the physical and social facilities constructed by municipal and national government agencies, and in turn businessmen were responsive to the urgings of the governor and local officials for private funding of new ventures and revitalization of cottage industries.

Finally, the social leaders of the city realized that changes in attitudes and behavior and in family and group relationships were an inevitable concomitant of economic and population growth in Chiangmai, and many sought ways of inducing change while preserving the most important cultural and social traditions. Some Buddist monks, for instance, took an active role in providing alternative social activities and outlets for young people when traditional temple activities fell out of favor during the 1960s. Many of the social elite, including members of the royal family, were active in organizing women's associations, child welfare programs, and charitable associations to assist the poor. Although many of the modern changes that came to Chiangmai were considered undesirable by some traditional groups in the city, many social leaders tried to accommodate and encourage, rather than suppress, change. Social leaders helped transform the social structure of the city to reinforce those influences promoting economic growth and widespread distribution of its benefits.[12]

EXPLOITATIONAL URBANIZATION: HUANCAYO, PERU

In his analysis of Huancayo, Roberts claims that the city's growth and development drained resources from its surrounding areas and benefited only the elite within the city and a small group of nonresident entrepreneurs.[13] Throughout most of its history, the city did little to transform its hinterland or to stimulate development among its rural population. Many of the factors that made Chiangmai a catalyst for regional development were lacking in Huancayo. Prominent landowners, merchants, and important commercial families who controlled economic activities in Huancayo had come from abroad or from other parts of the country, and did not have deep roots in the city. From early in Huancayo's history as a commercial and trading center, the merchant class dominated the political system and attempted to suppress or eliminate traditional social, political, and cultural activities of the peasants and working people who were born in the town. Traditional social activities such as fiestas and religious ceremonies were seen as frivolous and wasteful. Roberts notes that in the late 1800s the "Huancayo council concerned itself with the regulation and suppression of traditional fiestas; heavy fees were levied on those who wished to organize the traditional bullfights and dances. At times, these were forbidden entirely and fines were levied for infraction of the rules."[14]

Although Huancayo continued to grow as a commercial and trade center throughout the late nineteenth and early twentieth centuries, the increasing dominance of large-scale mining and plantation agricultural industries—which were operated to exploit local resources and invest the proceeds in Lima or abroad—changed the relationship of the city and its hinterlands in ways that were detrimental to the economies of both the city and the surrounding areas. Unlike the mutually beneficial ties that were forged by Chiangmai's industrial leaders with farmers in the countryside, in Huancayo there was virtually no relationship "between the major productive activities in the area—the mining and large-scale agriculture—and the patterns of local consumption." Roberts points out that "the products of mining and large-scale agriculture were mainly consumed in the industrial countries, and the manufactured goods consumed by the population of Huancayo and the area around were mainly produced abroad."[15]

The investments made by the city's elites and by nonresident entrepreneurs greatly changed the functional characteristics of the city during the early twentieth century by overwhelming the traditional market and trade activities that allowed Huancayo to grow from a small town to a service center for its rural hinterlands in the early 1800s. From the 1920s to the 1950s Huancayo took on more of the characteristics of an entrepot city—it facilitated the exploitation and export of raw materials from its hinterlands and channeled imported goods into the area, without contributing to productive capacity of indigenous economic activities or to the integration of the city with its rural hinterlands. Local shopkeepers became increasingly concerned with supplying the mining companies and less with serving local markets. Increasing pressures on the land and the growing involvement of the male labor force in mining left agriculture a low priority among many rural people and business leaders in the city. As a result, "crop agriculture did not significantly stimulate the development of locally-based commerce or industry. Flour mills failed because of uncertainties in the supply of wheat and even the brewery was to close partly because of the difficulties of obtaining a significant supply of barley."[16]

Moreover, the leading families of Huancayo came more and more to associate only with each other and to distinguish and separate themselves from the "common people" of the city. They established two social clubs to which only selected families and foreign residents could belong, and participated only perfunctorily in fiestas and traditional activities such as communal house building. There was little social interaction between the elite and workers and shopkeepers. The social services and facilities in which the elite invested tended to be reserved for their own use, a little of the benefit "trickled down" or spread to the larger population.

As the economy of the city came to be dominated by export production, many of the city's larger establishments fell under the control of Lima-based owners and managers. Few of the local agents who operated the establishments were committed to the economic growth of the city, and many of the government officials who were appointed to serve there saw it as a temporary position, a stepping-stone to an appointment in Lima. During the 1950s the local agents encouraged the immigration of large numbers of outside workers to take up employment in the mines and textile industries. Many of the local small-scale industries and shops could not grow because internal demand was limited by the low wages paid to local workers and because the meager salaries paid to immigrants were often sent to their families outside of Huancayo. Many of the influential economic and political leaders of the city had other, much larger, interests in Lima and were not overly concerned with obtaining national government resources for investment in Huancayo. Roberts observes that

> the growth of Huancayo was such that it did not become the location of any group—economic elite, "middle classes" or workers—that was sufficiently committed to the city and to its economic development to make the financial or personal investments that would have helped its longer term development. Not

only had the city's economy not contributed to the transformation of the countryside, but despite Huancayo's growth and economic diversification, it had not itself become transformed.[17]

The city became a place through which the constituent economic and social groups made their way to other activities in other places: Lima, their home villages, or other cities and towns. During much of the twentieth century the city "presented economic opportunities that could quickly be exploited but did not offer long term development that was worth fighting for, or conflicting over, with other groups."[18] When dominant economic activities declined, those associated with them moved on, and little attempt was made to readjust the economic, social, a political structures of the city, to diversify and transform its economic base, or to reintegrate the economic and social activities of the city with those of its rural hinterland.

ANTICIPATORY URBANIZATION: DAVAO CITY, THE PHILIPPINES

Hackenberg and Hackenberg have argued that even exploitational secondary cities, in the long run, may create conditions that can change the social and behavioral characteristics of urban and rural residents and provide opportunities for increased income and higher standards of living, through what they call "anticipatory urbanization."[19] In their studies of Davao City and its surrounding hinterland, they have noted that the economic base of the city, which was molded most forcefully during the American and Japanese occupations, and previously by a long period of Spanish colonialism, grew through the exploitation and export of local raw materials.

The social structure of the city reflected the colonial tradition well after the Philippines gained political independence. The timber industry, which for much of the twentieth century was the largest economic activity in Davao, was established under American, and fostered by Japanese, colonial regimes. After independence the industries were taken over by a few wealthy Filipino families, who operated them in much the same way as their colonial predecessors. The leadership of the city—dominated by these elite families who controlled large timber, agricultural plantation, and associated enterprises—had little personal concern with the city's social or economic welfare. As in Huancayo, the elite in Davao City still "belongs to a national (if not international) level social class with diversified family investments in plantation crops, urban real estate, transportation and mining interests throughout the country," Hackenberg and Hackenberg point out. "Descended from the old 'Spanish' landed nobility for the most part, it maintains a castle-like isolation and solidarity." These authors observe that "the name of its favorite local watering place, the Davao Insular Hotel, describes the behavior of the hotel's patrons as well as its location."[20] Absentee ownership of abaca, coconut, and rice plantations maintains the city's dual economic structure, for the plantations remain good investments for absentee owners only as long as they can exploit low-wage labor in the city and keep rural tenants in poverty.

The investment patterns of the timber industry owners clearly reflect their attitudes toward the future of the city. Lumber companies operating in the area have a choice of obtaining 2-year concessions to clear land, for which replanting and reinvestment requirements are minimal, or 25- to 50-year developmental concessions, which require the companies to restock the forested areas. During the 1970s, nearly 70 percent of the land in the Davao area was under 2-year licenses, and the timber resources have been depleted rapidly. Lumber industry owners are required by the government to build local plywood factories to ensure that processing and manufacturing activities are kept within the country, but most of the owners in Davao have sought government loans to meet this requirement rather than invest their own capital. Moreover, rather than linking their operations to local suppliers and processors in Davao City and its surrounding region, as did the lumber and tobacco industries in Chiangmai, the Filipino owners have internally integrated their investments in complementary activities, creating as little linkage as possible with other individuals or firms. Whenever possible, the timber companies have created their own plantations and shipping operations. The only benefits accruing to people living in Davao and its region are low-wage employment opportunities.

Hackenberg and Hackenberg note that

> absent from employment statistics and from the urban scene is any evidence of a major industrial payroll which large scale wood processing within the city would generate. Development economists who favor urban factory construction speak of the potential for backward and forward linkage to agricultural producers of raw materials. In Davao we seek in vain for the downward linkages to the industrial labor force.[21]

Thus, despite Davao's population growth and economic diversification, most of its residents and those of its rural hinterlands remain poor. The city has not greatly stimulated productivity and income in its region. Population growth is due mainly to continued immigration of peasants seeking low-wage employment.

Yet, even exploitational secondary cities can create conditions that are attractive to the rural poor and may, in the long run, provide opportunities for upward mobility and higher standards of living for a substantial number of people. More important, such cities can, slowly and indirectly, affect the attitudes and behavior of both the urban and rural poor so that, over time, as they inch up the social and economic ladder, they can change the direction and orientation of urban economic activities. Some evidence of anticipatory urbanization can already be seen in Davao. As lumber companies have cleared large tracts of land, these areas were subsequently settled by poor farmers, who, through cooperative arrangements, began to market surpluses and raise their incomes. The increased incomes stimulated greater demand for locally produced consumer goods, agricultural inputs, farm machinery, seeds, fertilizers, credit, transportation, and storage services. Moreover, since the primary markets for these surpluses were found in Davao City, new linkages of exchange were created with rural settlements, and their exposure to urban lifestyles and values has had a strong effect on farmers' methods of production and organization. "Anticipatory urbanization, acting as a creative force for change, will select those individuals most susceptible to modernization," Hackenberg and Hackenberg argue. "It will direct them, through migration, to form new agricultural communities which will be based on urban structural principles. The pre-urban communities will provide upward social mobility through commercial agriculture."[22]

These indirect and parallel opportunities for farmers seem to be matched by some degree of upward economic and social mobility that even exploitational cities can offer to the urban poor. Although, to outsiders, communities of squatters in Davao City may seem indistinguishable in their poverty, close observers point out that upward mobility is reflected in movement from one squatter settlement to another, and from one to other parts of the city. Even in exploitational secondary cities such as Davao, the timber industry, the "informal" or bazaar sectors, and small-scale service activities provide at least some income for most poor squatters. Feldman has observed in her studies of Davao's squatter settlements that "contrary to the sterotyped views of many upper class Filipino citizens concerning poor squatters, the latter do not live by begging or stealing, but by working."[23] In the 12 squatter settlements that Feldman studied in the 1970s, from 45 to 88 percent of the male household heads were regularly employed in at least half, and from 20 to 45 percent were regularly employed in the other half. Large percentages of male household heads were also employed casually, so that the average unemployment rate for all squatter barrios in Davao City during the 1970s was less than 8 percent. As the squatters, or their children, obtain larger incomes, they often move to what they consider to be better living conditions in slums in other parts of the city, and in the process they begin to acquire middle-class aspirations. Feldman's studies conclude that the "emergent middle class values of the Davao City squatters are evidenced by their concern for improved educational opportunities for their children, their desire to own their own homes even if the lots on which they reside are not theirs, and their concern for peace and order within their settlements."[24] For many, these aspirations are attainable—if not for themselves, then for their children.

Observers argue that under favorable conditions the incomes of farmers who settled land in the wake of timber clearing operations, and of the urban squatters who derive their income from services and employment in Davao City, can create increased demand for locally produced consumer goods and services. In the long run they could change the city's economic structure and its relationships with the rural hinterland. The children of these farmers and squatters, who are now seeking and obtaining education in the city, will be better prepared to acquire higher-paying jobs in the future or to engage in commercial activities that can satisfy the needs of a growing agrarian middle class and accelerate the pace of upward mobility within squatter communities. "This prospective development pattern contains the possibility of reorienting at least a part of the urban commerce away from the international raw materials markets and inward toward its own hinterland, no longer visualized as a resource base for extractive industries," Hackenberg and Hackenberg contend, "but as a source of increasing urban wealth through manufacturing and trade."[25]

It is to these possibilities, and to the factors that create a developmental pattern of urbanization in secondary cities such as Chiangmai, that national and international assistance efforts must be addressed. The historical factors that accounted for the growth of secondary cities, and that continue to stimulate the growth of smaller towns in developing countries, must be better understood. The dynamics of synergism, which maintains growth and diversification, should be encouraged and promoted. But population growth and economic diversification alone do not necessarily assure "developmental" urbanization. The factors that promote growth must be combined in ways that lead to widespread distribution of benefits, mutually beneficial linkages between urban and rural areas, and broad participation in productive activities. Even exploitational secondary cities offer some potential for development, however, if social and economic activities can be redirected. This requires building on the developmental functions, such as those identified in Chapter 4, that secondary cities can perform.

Policies for Secondary City Development

The challenge for international assistance organizations and national governments in the Third World is to find effective and appropriate ways to help local governments and private investors to strengthen the economies and service delivery capacities of secondary cities, both through direct investment and through national policies that have spatial implications. Assistance is required in two areas: first, in formulating national urbanization strategies for developing well-integrated systems of secondary cities; and, second, in helping city governments to solve the complex problems of managing urban growth.

Perhaps the most important role that international assistance organizations can play in those countries where national governments have committed themselves to pursuing secondary city development is to help formulate national urbanization strategies. An overall strategy for secondary city development is necessary because, as noted in Chapter 1, there are no universally applicable or optimal settlement patterns. Each country must attempt to shape its settlement system to meet its own national economic and social objectives, within its own resource constraints, and at pace with its own economic, administrative, and technological capacities. This study provides some evidence that secondary cities can play important roles in balancing the distribution of urban population and economic activities, in stimulating rural development, and in generating more socially and geographically equitable distributions of the benefits of urbanization when secondary urban centers are economically strong and linked to each other and to larger and smaller settlements within their regions.

Each developing country must fashion its own unique strategies for generating a strong, widely dispersed, and spatially integrated system of secondary cities. The objectives of promoting more balanced urbanization and greater equity in the distribution of benefits do not imply that all secondary cities must be developed simultaneously or that national resources must be distributed equally among them. Obviously, most developing countries have sufficient resources only to approach these objectives incrementally. Forging a more balanced urban system in the long run may require unequal distributions of resources and investments among cities and regions in the short run. The problems of polarized spatial development in most developing nations came about because of the skewed distribution of national investments, and their solutions may also require unequal distributions of national resources that favor secondary urban centers.

A variety of factors determine the extent to which resources can and should be focused on the development of secondary cities and the degree of deconcentrated urbanization that is feasible or desirable at any given time. Baldwin has identified and described some of those factors: the cost of urbanization, the amount of national resources available to influence the pace and direction of urbanization, the existing population distribution and density, the existing pattern of infrastructure development, the existing pattern of linkages among settlements, the degree to which economies of scale now exist in settlements, country size and topography, political feasibility, and administrative and technical capacity to promote different patterns of urbanization.[26]

In those countries where national governments are seeking to formulate strategies for secondary city development, three actions seem essential:

(1) strengthening the economies of existing secondary cities by: (a) extending basic social services and municipal facilities and infrastructure that support productive activities and improve human resources; (b) improving physical structure to make these cities more efficient and conducive to productive economic activities; (c) strengthening the economic base and employment structure; and (d) strengthening the planning, administrative, and financial capacity of local governments to manage urban development;

(2) Stimulating the growth and diversification of smaller towns and market centers to increase the number and geographic distribution of secondary cities within the national settlement system; and

(3) Strengthening the physical, economic, social, and political linkages among secondary cities and between them and larger and smaller settlements to provide greater access to urban services, facilities, and job opportunities to people living in rural areas,

and to create an integrated system of urban centers through which the benefits of urbanization and economic development can be spread more widely (see Figure 5.1).

STRENGTHENING EXISTING SECONDARY CITIES

Many of the specific ways to strengthen secondary cities will be discussed in Chapter 6. However, it should be noted here that promoting more balanced urbanization requires not only policies that meet the needs and solve the problems of secondary cities, but also complementary social and economic policies that are conducive to widespread urban growth. Current patterns of urbanization often reflect and are influenced by nonspatial development decisions. Policies on import substitution, trade regulations, foreign investments, migration control, monetary regulations, tax, wage, and price laws, land reform, and the structure of government organization may all, indirectly, affect the spatial pattern of economic development within a country and determine the degree to which secondary cities can compete with primate cities or national capitals.[27]

Moreover, it seems clear that policies must also be formulated to restrict or slow the expansion of the largest metropolitan areas and primate cities. Experience suggests that this is difficult to do until there are at least some secondary cities that are able to support high population threshold activities that would normally locate in the primate city. But countries such as South Korea have experimented with a number of programs for restricting the flow of migrants to the capital and redirecting people and educational, industrial, and commercial activities to secondary urban centers. The Korean government, over the past decade, has restricted the expansion of higher education institutions in Seoul, and is requiring branches of major universities to be located in cities outside of the capital's metropolitan area. It has restricted the construction of new high schools in the capital, provided funds

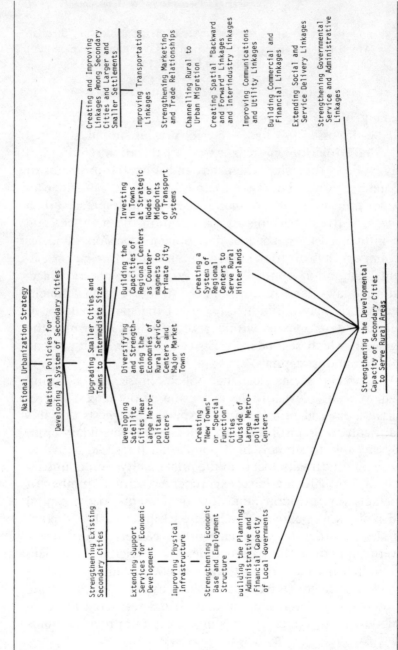

Figure 5.1 A Framework for Secondary City Development Strategy: Macro-Policies

National Urbanization Strategy

National Policies for
Developing A System of Secondary Cities

Strengthening Existing Secondary Cities

Extending Support Services for Economic Development

Improving Physical Infrastructure

Strengthening Economic Base and Employment Structure

building the Planning, Administrative and Financial Capacity of Local Governments

Upgrading Smaller Cities and Towns to Intermediate Size

Developing Satellite Cities Near Large Metropolitan Centers

Creating "New Towns" or "Special Function" Cities Outside of Large Metropolitan Centers

Diversifying and Strengthening the Economies of Rural Service Centers and Major Market Towns

Building the Capacities of Regional Centers as Countermagnets to Primate City

Investing in Towns at Strategic Nodes or Midpoints of Transport Systems

Creating a System of Regional Centers to Serve Rural Hinterlands

Strengthening the Developmental Capacity of Secondary Cities to Serve Rural Areas

Creating and Improving Linkages Among Secondary Cities and Larger and Smaller Settlements

Improving Transportation Linkages

Strengthening Marketing and Trade Relationships

Channelling Rural to Urban Migration

Creating Spatial "Backward and Forward" Linkages and Interindustry Linkages

Improving Communications and Utility Linkages

Building Commercial and Financial Linkages

Extending Social and Service Delivery Linkages

Strengthening Governmental Service and Administrative Linkages to increase educational services in regional centers, and made the transfer of high school students to Seoul more difficult. The Korean government has also taken actions that raise the costs or make it more difficult for large industries to locate in Seoul through zoning regulations, construction permits for factory building or expansion, and financial incentives for industrial relocation.[28] Although Seoul has continued to grow, its rate of growth during the past decade has been lower than in previous years because government policies have made location in the primate city more costly for both individuals and businesses.

The importance of having a deliberate and well-conceived national spatial strategy is reflected in the experience of both socialist and capitalist countries. The People's Republic of China's strategy, which will be described in more detail later in this chapter, has been to develop secondary cities as an integral part of national policy for promoting equitable growth, for strengthening the roles and functions of cities and towns of all sizes, and for developing rural areas. South Korea, through quite different methods, has attempted to do the same. Its industrialization policies have been aimed not only at increasing export production and output for domestic consumption, but also at generating and dispersing industrial employment and using small- and medium-scale industries to stabilize the economics and populations of rural areas. Its spatial policies have been integrated with economic policies to ensure that sufficient industrial land is available outside of the primate city to allow for a rapid increase in employment and production. They seek to disperse population from the two largest metropolitan centers and encourage a more balanced distribution of urban population, to create job opportunities in rural regions and foster development in economically lagging areas, and to achieve a "more rational allocation of industrial activities throughout the country."[29]

Although the details of policy implementation are still far from precise, the South Korean government's plans for balanced urban growth call for the creation of 8 planning regions in which 26 intermediate cities would be strengthened to serve both rural and urban residents. The country would be divided into 28 areas to be developed on the basis of their spatial and functional relationships, socioeconomic characteristics, and physical and economic comparative advantages:

(1) national metropolitan centers—Seoul and Pusan—in which national central management functions and highly specialized social and economic activities would continue to be concentrated;

(2) regional metropolitan centers—Taegu, Gwangju, and Daejeon—in which high-level commercial and administrative functions would be located and to which some of the population now migrating to the largest metropolitan centers might be attracted;

(3) urban growth centers—Chuncheon, Gangneung, Weonju, Chungju, Cheonan, Yeongju, Andong, Cheongju, Jeonju, Mogpo, Suncheon, Namwon, Jinju, Jeju, Pohang, Jeongju, and Jecheon—in which central place functions would be encouraged to serve both urban and rural residents; and

(4) rural service centers—Yeongweol, Hongseong, Gangjin, Geochang, Jeomchon, and Seosan—in which small-scale commercial, manufacturing, agroprocessing, and marketing activities would be strengthened to serve the rural population.

In addition, some cities—Incheon, Suweon, Anyang, Bucheon, Seongnam, Euijeongbu, Songtan, Gwangmyeong, Dongducheon, Masan, Changwon, Ulsan, Chungmu, Jinhae, and Kimhae—are to be developed as satellite centers for Seoul and Pusan and encouraged to perform supplementary social and economic functions for the largest metropolitan centers and to relieve some of the population pressures on their core areas.

The Korean urban strategy attempts to guide and direct national land development in a way that will "integrate large cities, medium, small cities and surrounding regions" within the settlement system. The country was divided into metropolitan and urban regions within which cities and towns would be classified and assigned special functions. The government earmarked special investments for each type of urban center. Transportation corridors were created to link metropolitan regional centers with each other and with Seoul by highway, rail, sea, and air, and by energy and fuel pipeline systems.[31]

Economic incentives and regulation were used to encourage the location of export industries in port cities, such as Incheon and Pusan. Heavy chemical, fertilizer, cement, and petroleum refining industries were encouraged to locate in government-created industrial estates in Pohang, Changwan, Ulsan, and Yonsan. Small and medium industries were encouraged to locate in other secondary cities such as Daejeon, Chuncheon, Jeongju, Mogpo, Gunsan, Donghae, and Weonju, which were provided with infrastructure, supporting services, and industrial estates. Financial incentives and other inducements were given to manufacturing firms to locate in secondary cities.[32] Moreover, during the 1960s and 1970s the Korean government used public investment in social overhead capital to increase the growth potential and employment capacity of intermediate cities. It invested heavily in electrical generating capacity, housing, and highway construction in and around selected inland cities and improved the cargo handling capacity and transportation access of coastal cities. It used existing secondary cities as a base for creating growth centers, rather than creating new growth poles. Thus, even though it somewhat mistakenly calls its strategy a "growth pole" policy, the Korean government is strengthening and developing a widely dispersed system of secondary urban centers.

The World Bank has suggested similar approaches in other developing countries and notes that building up the existing system of cities can be much more effective than trying to create a few new growth poles. The World Bank mission to Thailand suggested that the more evolutionary strategy of strengthening existing regional centers through "better exploitation of local entrepreneurship and productive resources" would also be more effective in achieving inter-regional equity than concentrating export production invest-ments in one or two industrial growth poles.[33] Bank analysts identified three essential elements of a strategy for secondary city development in Thailand:

(1) reducing the large gaps in services and infrastructure standards between the primate city—Bangkok—and other cities and towns within the country;

(2) altering industrial location incentives and related industrial promotion policies so that labor-intensive industries could locate in regional urban centers; and

(3) formulating and pursuing a coherent program of government investment in services and facilities to promote growth initially in those secondary cities with the strongest development poten-tial.[34]

The World Bank mission also suggested a policy of "domestic import substitution" to expand productive capacity within secondary cities. It called for a program to encourage manu-facturing of goods used in secondary cities or their rural hinterlands within the secondary urban centers, closer to their markets, rather than in Bangkok, where government incentives and subsidies for industry now attract nearly all large industries.[35]

Although little is yet known about the efficacy of national policies for changing urbanization and spatial development patterns, there are at least five actions that national and local governments can take to affect the economies of secondary cities. Friedly notes that these include:

(1) location of government's own facilities, offices, and enterprises in ways that attract related activities to particular cities or various sections or districts within cities;

(2) government controls on private sector office, business, or manufacturing activities through permits, taxes, and penalties;
(3) public inducements to private sector activities through subsidies, tax concessions, grants, and provision of required infrastructure to locate in specific cities or sections of cities;
(4) location of public services, facilities, infrastructure, and utilities that are needed by various economic activities in particular cities or sections of cities so that the costs of conducting businesses there will be lower than in other places; and
(5) indirect methods of government regulation on land use and costs, economic controls, and allocation of government expenditures to encourage or discourage development in particular cities or sections of cities.[36]

The problem of choosing which cities to favor through government action is, of course, a complex and politically sensitive one that must depend on careful analysis within each country. Although there are no universally applicable standards, Mosely has suggested some broad criteria that may be useful in choosing the initial cities for concentrated investment:

(1) location—existing or potential inter- and intraregional accessibility, and potential for becoming a service center for a wide area;
(2) human resources—the size, occupational range, quality, and diversity of the labor force, the quality of local leadership, entrepreneurial ability, and attitudes of government and business leaders toward local development;
(3) service capacity—the existence of, or potential capacity to provide, a wide range of services and facilities needed to attract industry, business, and professionals; and
(4) past growth performance—indications that the city already has some capacity to create new employment opportunities and to attract new business and industry.[37]

In formulating strategies for secondary city development, Mosely suggests that careful attention be paid to the distinction between active and reinforcing policies. The former seek to change the present pattern of development in secondary cities and to make them stronger catalysts for generating economic growth, while the latter seek to support and accelerate desirable changes already taking place.

These and other criteria for selecting secondary cities for development usually stress existing capacity and are most likely to favor larger metropolises that already have agglomeration economies and concentrations of services and facilities that will support more diversified economies. But, if the objective is to promote more balanced urbanization, then weaker secondary cities, smaller towns, and rural market centers must also eventually receive public investments in infrastructure and services, attract more diversified economic activities, and increase their population size.

UPGRADING SMALLER CITIES AND TOWNS

Cities of from 30,000 to 100,000 population that function as rural or regional service and commercial centers and rural market towns can, potentially, perform important roles in linking smaller settlements and intermediate cities and in spreading the benefits of urban development. With appropriate concentrations of investment these towns may be induced to perform more complex economic and social functions. Rondinelli and Ruddle have noted that such cities and towns already perform important development functions in many developing countries. They act as administrative centers for districts or provinces. They often support basic health services, including physicians, nurses, dentists, and maternity and health clinics. Usually they have an agricultural extension office, post office, primary and sometimes secondary schools, and a wide variety of small-scale commercial activities. In most towns and cities of this size the marketplace is the center of retail trade; it provides the major source of perishable foodstuffs and is the major outlet for surplus agricultural goods from the surrounding rural areas. The market is also the primary source of packaged, processed, and manufactured goods for rural consumers. Market towns and service centers are usually linked to larger cities by surfaced roads in all but the most remote regions of developing countries, and are connected to smaller towns and villages by access roads.[38] Investments in service centers and market towns should be aimed at strengthening their existing functions and increasing their capacity to perform a wider range of functions and offer a greater variety of services and commodities than they usually do. A diversified set of enterprises must be encouraged to grow in market towns so that the benefits of association and proximity can provide economies of scale that allow them to thrive and to attract still other economic activities. As the number of trading, manufacturing, and service industries grow in one center, there is a strong probability that the total demand for all services will increase and that the towns' service areas will expand.

Both secondary cities and smaller service and market towns that are chosen for concentrated investment by the national government should be strategic locations that will contribute to regional development. Richardson has suggested four ways of building up the system of secondary cities:

(1) promoting the growth of small and middle-sized cities that are close enough to major metropolitan centers to benefit from their agglomeration advantages, yet that are not so close that they are "swallowed up" in the growth and expansion of the metropolis;

(2) promoting the growth of cities and market towns far away from the largest metropolises or the primate city, as a countermagnet to attract rural migrants and high-threshold economic activities that currently locate in the primate city;

(3) developing smaller cities in underdeveloped and sparsely populated rural regions so that they begin to generate agglomeration economies; and

(4) developing transportation axes that connect existing and potential secondary cities and that create conditions conducive to the growth of multiple midpoint centers or nodal centers at terminal points or "breaks" in the transportation network.[39]

Methodologies already exist for analyzing national and regional spatial systems to identify settlements that might be developed or upgraded. USAID has tested a macro-spatial analysis methodology through its "Urban Functions in Rural Development" projects in the Philippines, Bolivia, the Cameroons, Upper Volta, and a few other countries, for instance, and similar methods of analysis have been used by the Ford Foundation in India and Indonesia for determining the characteristics of the existing hierarchy of settlements and for gauging the strengths of linkages among secondary and smaller cities. USAID's methodology involves ten phases:

(1) analysis of the demographic, social, economic, and physical characteristics of the region under study that serves as a data inventory for planning and as a baseline study for monitoring and evaluating changes in the settlement system;

(2) analysis of the existing spatial structure that describes elements of the settlement system, the functional complexity and centrality of settlements, the hierarchy of central places, and the distribution and patterns of association among functions within the study area;

(3) description and analysis of the major socioeconomic, organizational, and physical linkages among settlements within the study area and between them and settlements located in other regions of the country;

(4) mapping of information obtained from the functional complexity, settlement hierarchy, and spatial linkage analyses to determine the "areas of influence" or service areas of settlements in various functional categories within the study area;

(5) delineation of areas where linkages are weak or nonexistent, and of marginal areas that are not served by central places, or

in which the rural population has poor access to town-based services and facilities that are crucial for rural development;

(6) comparison of information from the demographic, socio-economic, and physical surveys and the settlement system, functional distribution, and linkage analyses to national and regional development plans and objectives to (a) determine the adequacy of the settlement system to meet development needs and implement equitable growth policies and (b) identify major "gaps" in the settlement system, in service areas for important functions and in linkages among subareas of the region or country;

(7) translation of the spatial analyses into an investment plan that identifies the projects and programs that will be needed to ameliorate major development problems, to strengthen and articulate the regional spatial structure, and to integrate various levels of settlement within it;

(8) integration of projects identified through spatial and economic analyses into coordinated "investment packages" for different locations within the region or country, and the combination of the investments into a priority-ranked and appropriately sequenced investment budget for the development of various subareas or regions over a given period of time;

(9) creation of an evaluation system for monitoring the implementation of projects and programs and for determining the substantive results of development activities on marginal areas and population groups within selected regions or subareas of the country; and

(10) institutionalization of the planning procedures in local and regional public agencies charged with investment decision making and with revising the spatial analysis and development plans at appropriate intervals.[40]

This analytical approach can be used to identify cities and towns with functional characteristics that might be upgraded or strengthened, and as a framework for more detailed studies of particular towns and cities before investment programs are designed.

CREATING AND IMPROVING LINKAGES
AMONG SECONDARY CITIES AND
BETWEEN THEM AND OTHER SETTLEMENTS

Strengthening the linkages among settlements is an essential part of a national strategy for building the capacity of cities to perform their developmental functions more effectively because, as was noted in Chapter 5, unless secondary cities are linked to smaller and larger places it is unlikely that they will play a catalytic role in stimulating regional development. Linkages are important for a number of reasons. Richardson points out that "paradoxically, the most effective methods of remedying a deficiency of medium-sized towns may not be to stimulate them directly but to capitalize on the systematic interrelationships of the hierarchy by creating the conditions under which they may develop spontaneously."[41] As was noted in Chapter 4, historically, the growth of secondary cities depended strongly on the creation of transport, trade, administrative, social, and physical linkages that stimulated economic diversification and attracted sufficient numbers of people and activities to generate economies of scale.

Contemporary studies also indicate that increasing the linkages among settlements stimulates the growth of new centers and that the growth of existing centers creates new linkages among them. World Bank analysts conclude from their study of regional urban centers in Thailand that "one of the strongest reasons for promoting regional cities is to strengthen urban-rural linkages with the objective of retaining as much population as possible in the rural areas by expanding opportunities."[42] These researchers conclude that "although rural development is critically important in Thailand, these regions cannot be developed by rural measures alone." They argue that the development of regional cities would "create employment opportunities for surplus rural population and, if the linkages between the cities and the rural areas are carefully forged, help improve the stability and vigor of the rural economies."[43]

Analyses of market towns and urban marketplaces in the Philippines have concluded that the financial feasibility of public markets depended on volume of sales and that those public markets that were not operating efficiently tended to have weak links to their rural hinterlands. The studies conclude that "projects which link urban areas with their rural hinterland may have an important positive effect on public markets, especially for large markets."[44]

The International Labor Organization (ILO) team that sought to fashion a full employment strategy for Colombia has pointed out that although extending services and facilities in secondary cities is often more costly than in the largest metropolises, "all policies designed to redress regional and rural-urban imbalances must give top priority to the reduction and eventual elimination of differences in quality between supposedly equal services."[45] ILO analysts note that in the initial stages of investment, costs of extending basic services to secondary and smaller cities can be lowered by using a hierarchy of services and facilities that is related to the hierarchy of settlements. Higher-order, more costly, higher-threshold services can be located in larger settlements that draw on greater service areas and populations; lower-order, less expensive, less sophisticated services can be located in smaller settlements. The strategy will work well, however, only if the components are well integrated and linked through the settlement system. "A health post lacking adequate means of communication with the nearest health center or hospital—a telephone, for example—cannot possibly fulfill its functions properly in a system where the possibility of referral is vital," the ILO mission emphasizes.[46] Different types and grades of roads must be linked to each other, communications and energy systems must link smaller and larger cities, and rural services and facilities must be linked to those in towns and cities at each higher level in the urban system. These systems must be developed at pace with the extension of services and with the provision of incentives to strengthen the economies of secondary cities. The way institutions are organized and the kinds of organizations established in secondary cities can also play an important role in forging linkages among them and with rural towns and villages. Deliberate attempts to create contracting arrangements between large commercial and industrial establishments in big cities, and smaller agricultural and manufacturing suppliers in smaller cities and towns, for example, can integrate and link urban and rural economies into a national system.[47]

A number of the case studies reviewed here point out that secondary cities can have a pervasive impact on their regions by diffusing new ideas, creating economic opportunities for people living in nearby towns and villages, extending services into their peripheries, generating new employment, and integrating towns and villages in surrounding rural areas into the regional economy. The degree to which any particular city performs these functions, however, is not easy to determine and, as noted in Chapter 3, not all secondary cities perform regional integration and development functions well. Some cities have not been catalysts for development, but instead have exploited the resources of their regions to promote their own growth and development. But most urban economies have some stimulative aspects and they must be better understood in order to reinforce and expand those activities that will generate higher levels of productivity and income for people living in secondary urban centers and their rural hinterlands.

To a large extent, the diffusion and integration functions of secondary cities are performed through their other roles as centers of public and social services, as central places for the provision of commercial and personal services, as centers of manufacturing, transport, and communications, as absorbers of rural migrants, as centers of social transformation, and especially as regional marketing, agroprocessing, and trade centers. Some studies have shown that the trade and exchange functions of secondary cities are among the strongest forces for stimulating economic development and integrating urban and rural settlements within regions, in both developing and Western countries.[48] Skinner found that early in China's history, larger towns and cities with regional marketing functions integrated rural areas within their fields of market influence to form a well-ordered and highly articulated hierarchy of settlements through which people living in the countryside had easy access to economic and social functions supported by towns and cities of various sizes. Intermediate cities encompassed smaller market areas, supplied villages and towns with many goods and services, and provided outlets for rural products. Intermediate cities had two service areas—those nearby villages from which people attended the urban market regularly and more distant villages from which people came periodically, or to which itinerant merchants went to collect products for resale in the city. Skinner found that "an intermediate market town functions as the nucleus not only of the larger intermediate market system but also of a smaller marketing system."[49]

The case histories of many of the contemporary secondary cities reviewed here also refer to their impact in diffusing innovation, linking rural villages and towns to the urban economy, and integrating urban and rural sectors through a variety of administrative, social, economic, and political interactions. But the difficulty of determining the degree to which secondary cities actually perform these functions stems in part from the fact that as cities grow they tend to become more independent of their regions in some respects as well as more dependent on them in others. This mixture of regional integration and autonomy is clearly noted in Hazelhurst's description of middle-range cities in India and is most

apparent in the patterns of marketing and business trans-
action, in which merchants and traders have different sets of
criteria for doing business with local residents and "outsiders."[50] The large role that markets play in the economy of
secondary metropolises is evidence of extensive interaction
between city and countryside. "In India, regional integration
of the middle city with the surrounding countryside is maintained by the bazaar economy, through which relationships
are established between shopkeepers and their clientele,"
Hazelhurst points out. "A second level of regional integration
is maintained by collateral relationships among merchants
within a city and between merchants in neighboring
cities."[51]

In many African secondary cities, urban brokers, distributors, and traders play an important role in linking rural and
urban economies. Trager observes that in Ilesha, Nigeria,
urban middlemen play an important role in bulking agricultural products in village markets and on individual farms.
They resell some products to other intermediaries for redis-
tribution to more distant cities and towns, and they buy
manufactured goods in larger cities such as Ibadan and Lagos
and sell them to traders in Ilesha and to farmers in smaller
towns and villages in Ilesha's trading area, as well as in the
Ilesha market.[52] The distance of trade ranges from less than
ten miles for local goods to several hundred miles for manufactured products. In many African and other developing
countries, urban middlemen not only trade products, but
information and gossip as well; they spread the word about
new lifestyles and new ways of farming, and form a human
communications network that is sometimes more efficient
than a technological system.

As most secondary cities in Africa have grown larger, their
marketing areas have expanded and their linkages with other
cities and towns have become more extensive. Jones notes
that as Ibadan grew into a secondary city its "staple supply

hinterland wound around and leaped over the supply hinterland of neighboring cities," encompassing them in much the same way that larger markets encompassed smaller ones in China and in many Western nations. Before the growth of Ibadan, the market areas of smaller cities such as Ilorin and Ilesha were "arranged like tiles across the landscape with each little city surrounded by its farmlands, from which came its basic food supply." But when Ibadan's growth created demands that were beyond the ability of its hinterlands to meet, the city's merchants reached out to other cities and their hinterlands to obtain supplies and the marketing interactions drew these other cities and rural areas into Ibadan's area of market influence.[53]

In more urbanized developing countries, the growth of secondary cities tends not only to integrate towns and villages into the urban system, but to articulate the settlement system as well—that is, to promote the growth of smaller towns into larger ones capable of supporting more and higher-order functions—and tends to generate a rank ordering of towns in a hierarchical fashion in the area surrounding the secondary city. Ajaegbu observed this articulation and integration of settlements in areas surrounding secondary cities in Nigeria and noted that the relationships among centers and between them and their surrounding towns and villages "are very complex and dynamic." He observes that as larger numbers of towns grow, fuctions are more widely dispersed throughout a region and throughout the country. "More centres are today being looked upon for functions of the high or highest order than hitherto," he states. "More subgrouping of settlements in the natural hierarchy has occurred as a result of the various levels of deconcentration of central place functions."[54]

A similar process seems to have occurred in some developing countries in Asia, notably Taiwan and South Korea, where urbanization and industrialization have been rapid during the past quarter century. Moreover, the spread of modern services and facilities from the secondary cities seems to have increased the productivity, income, and living standards of people residing in rural areas immediately surrounding the cities. Park's analysis of the rural areas surrounding Taegu, Korea, during its period of rapid growth during the 1960s concludes that rural roads, educational facilities, public services, and utilities were far better in those towns and villages than in areas more remote from the city. Farmers in nearby areas had greater access to transportation, farm machinery, and agricultural supplies and equipment.[55] Moreover, the city provided employment for rural youth from nearby villages, allowing them to supplement their incomes and become socialized into urban life and yet remain in their rural homes and participate in farm work during peak periods of agricultural activity.[56]

The dissemination of information, innovation, and urban attitudes and behavior takes place not only through these mechanisms, but through others as well. A study by Collier and Lal for the World Bank found that in Kenya, employment of rural migrants in cities not only provided a steady stream of remittances to rural villages, but explained the relatively high level of agricultural innovation among smallholders. The remittances were a major source of funding for the adoption and use of new agricultural technology. "It appears that educated rural-urban migrants with formal sector jobs are the major source of urban-based off-farm income, which in turn is the major determinant of the levels of smallholder innovation," these analysts conclude. "Thus, the faster urban formal sector employment grows, the greater are the urban based non-farm income streams and hence the faster the spread of innovation among smallholders."[57]

Mortimore's studies of the rural villages and towns around Kano, Nigeria, also indicate the degree to which secondary cities can intensify commercial agriculture, increase nonagricultural occupations in nearby villages, raise off-farm income, and stimulate cottage industry. Population density and trade relationships with Kano became stronger for villages in the "close-settled zone" around the city as its area of economic influence expanded. Moreover, weaving and other cottage industries that provided goods for the market in Kano tended to make up for any deficiencies in agricultural output that may have occurred from the shift in occupational structure in the villages. "The close ties between town and country," Mortimore notes, "has played an essential part in the growth of the zone."[58]

Although most case studies highlight the stimulating influence of urban growth on areas surrounding the city, in some regions agricultural and rural development have stimulated urban economic growth. The forces for regional integration came from rural areas. This seems to have been the case in Meerut City, in India, where Sundaram found that rising productivity in rural areas of Meerut District due to the application of "Green Revolution" technology created demand for farms supplies and equipment produced in the city, generated more trips by farmers to the urban center, and increased the frequency of visits for shopping, business, and entertainment, thereby forging "closer economic relations of the rural sector with the urban sector."[59] Moreover, the growing prosperity of the countryside led to diversification within villages, encouraged rural industrialization, and "created the need for urban markets, raw-materials, know-how and skilled manpower and thus had led to the closer interaction of the rural areas with the city." In addition, the extension of social infrastructure and public services into rural areas from Meerut City, according to Sundaram, "has strengthened the linkages with urban institutions concerned with health, education and culture and has increased rural-urban interaction" in service as well as commercial activities.[60]

Limited Spread Effects:
The Need for a Network of Secondary Cities

Although secondary cities can have strong and pervasive impacts on the development of their regions, the areas of influence are clearly limited and the impact of the city on villages and towns declines with distance, depending on its size and economic diversity. From his studies of diffusion influences in Latin America, Stohr suggests that in early stages of urbanization and development, when communications and transport systems are weak, diffusion of innovation depends less on city size relationships than on physical distance among settlements. Those places nearer secondary cities of any size are likely to be exposed to new ideas and methods more quickly than remote settlements.[61] But the extent of the "spread effects" of development from secondary cities has been the subject of debate among planners and development analysts over the past few years, and there are still relatively few data on which to base firm conclusions.

Using nearly two dozen socioeconomic variables in the analysis of the incidence of development in the vicinity of Medellín, Colombia, Gilbert found that in the mid-1960s development scores were higher for communities within a 25-km band around the city than elsewhere in the province and dropped sharply for towns more than 50 km from Medellín. Gilbert notes that in every case, "the *municipios* within 50 kms of Medellín showed higher values than those outside of this band and varied in direct relation to their distance from the centre within the 50 km circle. Outside the band there was a gradual decline with distance up to 150 kms and in certain cases beyond."[62] Gilbert tentatively concluded, as have other analysts, that social services and infrastructure improvements in larger urban centers do diffuse throughout the immediate areas surrounding them, but that they become relatively weak in areas farther away from the city.

One conclusion that can be drawn from this and other analyses of the impact of secondary cities is that the creation of isolated industrial "growth poles" in rural regions of a developing country is not sufficient to stimulate widespread economic growth in rural areas or to spread the benefits of urbanization equitably throughout a developing country. Because the spread effects tend to weaken rather rapidly with distance, a system of secondary cities connected to smaller cities and towns—which in turn are linked to rural villages and farm areas—seems necessary to ensure the diffusion of innovation, the integration of urban and rural areas, and the stimulation of economic activities within a region.[63] Moreover, simply creating a system of secondary cities alone does not seem to be sufficient to achieve the benefits of urbanization. Stohr notes in his analyses of Latin America that the existence of smaller cities and towns to which innovations, capital investment, services, facilities, and productive opportunities can be spread seems to be a condition for making secondary cities more developmental. He argues that

> if an intermediate-sized (and fast growing) city exists in a region but no lower ranking cities exist (e.g. a mining town in an otherwise little developed resource frontier), the city will become an enclave and develop at the expense of the rest of the region. It therefore seems that, when talking about growth centres, we are dealing with a chain process where the actual efficiency of each link is determined not only by its own strength but by that of its superior links (for supplying impulses) and by that of its subordinate links (for receiving impulses).[64]

Misra and Sundaram come to similar conclusions in their analysis of industrial growth poles in India; they claim that in many, the spread effects of concentrated industrial development have been narrowly constrained because the linkages between the growth poles and rural areas and towns in their regions are weak. "What is clear," they conclude, "is the fact that unless the new growth centres are planned as regional centres capable of serving the region they are located in, they cannot become instruments of modernization."[65] It is this combination of internal and external linkages that seems to make secondary cities catalysts for development. Stohr argues that the only cities in Latin America that have been able to act effectively as *regional* growth centers are those that have developed a combination of externally oriented *and* regionally based economic activities, characteristics that, until recently, only coastal cities and national capitals have acquired in most Latin American countries. These cities, "while producing for extra-regional (national or international) demand, usually possess sufficient integration between regional supply factors (capital, technology, labour, societal innovation) and regional demand (effective purchasing power) to provide for self-sustained growth."[66]

Although the experience in countries that have attempted to promote widespread urbanization, decentralize industrialization, and build on the agricultural, service, commercial, and trade functions of secondary cities has not been systematically analyzed—largely because the experiments have been so recent—there is some evidence that the emergence of a system of secondary cities can have a positive effect on both reducing the polarization of the urban settlement system and stimulating development in rural areas. Much of the initiative must come from national governments with investment resources to build up the capacities of secondary cities to serve their regions and to meet external demand. But a good deal of the effort must also come from within secondary cities, and that initiative depends on the strength of political will to create more balanced economic growth within a region and to use the economic resources of the city to stimulate development in the countryside. As noted earlier, many of the factors that affect the degree to which secondary cities become developmental depends on the attitudes and actions of public and business leaders within the city. This is confirmed in one of the few studies of a region in which political forces affecting a city's impact on its region were examined in detail. Spodek found in his study of Saurashtra, India, that "under appropriate circumstances the city could function as market place and growth pole, but only when the dominant political interests wished it to." He concludes that in this region the cities "became most productive economically after independence as a new interest group, desiring balanced economic growth, came to power. The urban sector served the political interests which dominated it and regulated its character."[67]

Similar conclusions can be drawn from Taiwan's experience during the 1960s and from Korea's during the 1970s.[68] These lessons seem to have been learned well by the government of the People's Republic of China during the 1960s and 1970s. During the period when its highest priorities were to attain widespread and geographically equitable development, policies were aimed at carefully allocating investments among cities of different sizes to promote more widespread distribution of income, reduce regional disparities, and create a hierarchy of settlements capable of performing different functions within each region and throughout the country.[69] The Chinese used three levels of municipalities—provincial capitals, prefecture capitals, and rural towns or *hsien* cities to decentralize economic activities. The provincial capitals received the bulk of investments for heavy industry—iron, steel, machinery, textiles, motor vehicles, and farm equipment—modern infrastructure, heavy utilities, and major highways. Prefectural cities received funds for investments in light industry, agroprocessing, simple machine tools, electric motors, and light farm machinery, and those manufacturing activities using intermediate technology and locally available materials. The *hsien* cities were conceived of as centers of direct urban-rural interaction and were designed to provide small components for manufacturing establishments in larger cities and farm inputs to rural villages. *Hsien* cities were made responsible primarily for the production of energy, cement, fertilizer, iron, and simple farm implements, for repair of agricultural equipment, and other production functions that could be performed by small-scale units.

In all cases, the Chinese reinforced the existing system of cities rather than creating "new towns." Chang observes that "the overwhelming majority of the *hsien* cities today were walled cities in imperial times and the present regime has simply transformed the old administrative centers into local industrial production centers, thus enhancing and endowing their traditional urban network with modern technology."[70] As smaller cities grew in population they were upgraded in classification from small to intermediate cities and allocated new functions. One analyst estimates that from 1953 to 1972 the number of cities with from 50,000 to 100,000 population increased from 71 to 105, and those with from 100,000 to 500,000 increased from 77 to 91.[71]

Perhaps the most important factors in China's apparent success with achieving more balanced urbanization were that smaller secondary cities were closely linked to activities in their rural hinterlands and that their industrial functions were run mainly to increase the productivity of rural people. China attained a surprisingly uniform distribution of secondary cities by the early 1970s, given the diversity among regions in natural resource endowments, physical conditions, and economic capability. As a result of its development strategy, a base was established for redistributing urban population and productive activities. Population growth was slowed in the largest cities while it increased in intermediate and smaller urban centers. Chang predicts that as a result of these policies "the largest cities will likely increase at the slowest rate, while the most rapid growth rate should occur in settlements with a population under 50,000, which in many ways have closer ties to the modernization process in rural areas."[72]

Although little is known about spatial development policies in China since the death of Mao, there is nothing inherently contradictory between the accelerated modernization policies of the new regimes and the spatial system that was shaped during the period prior to 1972. Indeed, if equitable growth is still an important goal, and a great deal of evidence suggests that is is, the network of secondary cities that was built during the 1950s and 1960s should provide a broad base for industrialization and modernization in China in the future.[73]

Under proper conditions secondary cities can be forces for the development of their hinterlands and for integrating the economies of developing regions. Creation of industrial growth poles in the ways attempted by many countries during the 1960s, however, seems to be neither appropriate nor sufficient to generate widespread development. The service, distribution, commercial, marketing, agroprocessing, and other functions of secondary cities may offer a far better base for stimulating their growth than large-scale manufacturing. And even if industrialization is the basis for secondary city development, it is clear that the economic activities encouraged within secondary cities must create and serve regional demand as well as external markets. Moreover, because of the limited distance over which the "spread effects" of urban economic and social activities seem to distribute benefits, isolated growth poles within rural regions are unlikely to have much impact. A well-distributed system of secondary cities seems to be needed to ensure that the benefits of urban growth spread to towns and villages in rural regions. Secondary cities, in turn, must be linked to smaller cities and market centers for innovations and economic stimulants to make their way "down" the hierarchy of settlements and "outward" from the urban core. Unless they were linked to both larger and smaller settlements, the industrial growth poles created during the 1960s and 1970s simply became enclaves of urban development without generating benefits for their regions. The creation of a system of secondary cities may allow these larger urban centers to increase the access of rural people to their urban services and facilities, job opportunities, and amenities more effectively. Whether secondary cities grow spontaneously or their growth is stimulated by government investment, their development creates new problems as well as new opportunities, and both must be considered in international assistance programs and national development strategies.

NOTES

1. U.S. Agency for International Development, *Colombia: Urban-Regional Sector Analysis* (Bogota: Author, 1972, p. 13.

2. Ibid.

3. Luis Unikel, "Urbanization in Mexico: Policies, Implications and Prospects," in *Patterns of Urbanization: Comparative Country Studies,* vol. 2, ed. S. Goldstein and D. F. Sly (Liege, Belgium: International Union for the Scientific Study of Population, 1977), p. 538.

4. Ibid., p. 539.

5. Russell J. Cheetham and Edward K. Hawkins, *The Philippines: Priorities and Prospects for Development* (Washington, DC: World Bank, 1976), p. 306.

6. Evangeline P. Javier, "Economic, Demographic and Political Determinants of Regional Allocation of Government Infrastructure Expenditure in the Philippines," *Journal of Philippine Development* 3, no. 2 (1976): 281-312.

7. See Dennis A. Rondinelli, "Regional Disparities and Investment Allocation Policies in the Philippines: Spatial Dimensions of Poverty in a Developing Country," *Canadian Journal of Development Studies* 1, no. 2 (1980): 262-287; E. B. Prantilla, *Industrialization Strategy and Growth Pole Approach to Regional Development* (Nagoya, Japan: United Nations Centre for Regional Development, 1972), pp. 19-22.

8. See Michael A. Cohen et al., *Urban Growth and Economic Development in the Sahel,* World Bank Staff Working Paper 315 (Washington, DC: World Bank, 1979), p. 13.

9. Chakrit Noranitipadungkarn and A. Clarke Hagensick, *Modernizing Chiengmai: A Study of Community Elites in Urban Development* (Bangkok: National Institute of Development Administration, 1973).

10. Ibid., p. 29.

11. Ibid., p. 83.

12. Ibid.

13. Bryan Roberts, "The Social History of a Provincial Town: Huancayo, 1890-1972," in *Social and Economic Change in Modern Peru,* ed. R. Miller, C. T. Smith, and J. Fisher (Liverpool: Centre for Latin American Studies, University of Liverpool, 1976), pp. 136-197.

14. Ibid., p. 148.

15. Ibid., p. 150.

16. Ibid., p. 151.

17. Ibid., p. 168.

18. Ibid.

19. Robert A. Hackenberg and Beverly H. Hackenberg, "Secondary Development and Anticipatory Urbanization in Davao, Mindanao," *Pacific Viewpoint* 12, no. 1 (1971): 1-20.

20. Ibid., p. 8.

21. Ibid., p. 12.

22. Ibid., p. 4.

23. Kerry Feldman, "Squatter Migration Dynamics in Davao City, Philippines," *Urban Anthropology* 4 (Summer 1975): 134.

24. Ibid., p. 142.

25. Hackenberg and Hackenberg, "Secondary Development and Anticipatory Urbanization," p. 15.

26. See Emily Baldwwn, *The Role of Urban Decentralization in Developing Nations: Toward Policy-Development Guidelines* (Washington, DC: U.S. Agency for International Development, 1980), pp. 19-29.

27. J. T. Fawcett, R. J. Fuchs, R. Hackenberg, K. Salih, and P. C. Smith, *Intermediate Cities in Asia Meeting: Summary Report* (Honolulu: East-West Center Population Institute, 1980), pp. 9-16.

28. See Son-Ung Kim and Peter J. Donaldson, "Dealing with Seoul's Population Growth: Government Plans and Their Implementation," *Asian Survey* 19 (July 1979): 660-673.

29. Korea Research Institute for Human Settlements, *National Land Development Planning in Korea* (Seoul: Author, 1980), p. 17.

30. Ibid., pp. 38-57; see also Republic of Korea, [Long Range Planning for Urban Growth to the Year 2000] (unofficial trans.) (Seoul: Ministry of Construction, 1980), pp. 137-143.

31. Korea Research Institute for Human Settlements, *National Land Development,* pp. 17-18; see also *Dong-A Ilbo,* August 24, 1981, which provides a detailed account of the Second Long Range National Land Plan.

32. See Korea Research Institute for Human Settlements, *National Land Development,* pp. 17-18; Byung-In Park, "Analysis of Urban Growth for the 1970s," in [Municipal Affairs] (Seoul: Korea Local Administration Association, 1979), p. 49.

33. Frederick Temple et al., *The Development of Regional Cities in Thailand* (Washington, DC: World Bank, 1980), p. 19.

34. World Bank, *Thailand: Urban Sector Review,* Background Working Paper 7 (Washington, DC: Author, 1978), p. 39.

35. Ibid., p. 43.

36. Philip H. Friedly, *National Policy Responses to Urban Growth* (London: Saxon House, 1974), pp. 123-125.

37. M. J. Mosely, *Growth Centers in Spatial Planning* (Oxford: Pergamon, 1974), pp. 158-173.

38. Dennis A. Rondinelli and Kenneth Ruddle, *Urbanization and Rural Development: A Spatial Policy for Equitable Growth* (New York: Praeger, 1978), ch. 3.

39. Harry W. Richardson, *City Size and National Spatial Strategies in Developing Countries,* World Bank Staff Working Paper 252 (Washington, DC: World Bank, 1977), pp. 53-54.

40. Dennis A. Rondinelli, *Spatial Analysis for Regional Development: A Case Study in the Bicol River Basin of the Philippines* (Tokyo: United Nations University, 1980), pp. 21-40.

41. Richardson, *City Size and National Spatial Strategies,* p. 54.

42. Temple et al., *Development of Regional Cities,* p. 20.

43. Ibid., p. 21.

44. Roy Bahl et al., *Strengthening the Fiscal Performance of Philippine Local Governments* (Syracuse, NY: Syracuse University Local Revenue Administration Project, 1981), pp. ES33-ES34.

45. International Labor Organization, *Towards Full Employment: A Programme for Colombia* (Geneva: Author, 1970), p. 100.

46. Ibid., p. 101.

47. See M. I. Logan, "Key Elements and Linkages in the National System: A Focus on Regional Planning in Nigeria," in *Planning for Nigeria: A Geographical Approach,* ed. K. M. Barbour (Ibadan: Ibadan University Press, 1972), pp. 16-39.

48. See especially E.A.J. Johnson, *The Organization of Space in Developing Countries* (Cambridge, MA: Harvard University Press, 1970); similar arguments are made by Dennis A. Rondinelli and Kenneth Ruddle, *Urbanization and Rural Development: Spatial Policies for Equitable Growth* (New York: Praeger, 1978).

49. G. William Skinner, "Marketing and Social Structure in Rural China, Part I," *Journal of Asian Studies* 24 (November 1964): 24.

50. Leighton W. Hazelhurst, "The Middle Range City in India," *Asian Survey* 8, no. 7 (1968): 540-542.

51. Ibid., p. 547.

52. Lillian Trager, "Market Women in the Urban Economy: The Role of Yoruba Intermediaries in a Medium Sized City," *African Urban Notes* 2, no. 3 (1976-1977): 6.

53. William O. Jones, "Some Economic Dimensions of Agricultural Marketing Research," in *Regional Analysis,* vol. 1, ed. Carol A. Smith (New York: Academic, 1976), pp. 313-314.

54. H. I. Ajaegbu, *Urban and Rural Development in Nigeria* (London: Heinemann, 1976), p. 56.

55. Park, "Analysis of Urban Growth," pp. 145-147.

56. Ibid., pp. 131-132.

57. Paul Collier and Deepak Lal, *Poverty and Growth in Kenya,* World Bank Staff Working Paper 389 (Washington, DC: World Bank, 1980), p. 43.

58. M. J. Mortimore, "Population Densities and Rural Economies in the Kano Close Settled Zone, Nigeria," in *Geography and the Crowding World,* ed. W. Zelinsky, L. A. Kosinski, and R. M. Prothero (New York: Oxford University Press, 1970), p. 389.

59. K. V. Sundaram, *Role of Cities in Attaining a Desirable Population Distribution in the Context of Rapid Urbanization: Case Study of Meerut, India,* Research Project 501 (Nagoya, Japan: United Nations Centre for Regional Development, 1975), p. 32.

60. Ibid.

61. Walter B. Stohr, "Some Hypotheses on the Role of Secondary Growth Centres as Agents for the Spatial Transmission of Development in Newly Developing Countries–The Case of Latin America," in *Proceedings of the Commission on Regional Aspects of Development of the International Geographical Union,* vol. II, ed. F. Helleiner and W. Stohr (Ontario: International Geographical Union, 1974), pp. 75-111.

62. See Alan Gilbert, "A Note on the Incidence of Development in the Vicinity of a Growth Centre," *Regional Studies* 9 (1975): 325-333.

63. An approach is outlined in Rondinelli, *Spatial Analysis for Regional Development.*

64. Stohr, "Some Hypotheses on the Role of Secondary Growth Centres," pp. 98-99.

65. R. P. Misra and K. V. Sundaram, "Growth Foci as Instruments of Modernization in India," in *Regional Policies in Nigeria, India and Brazil,* ed. A. Kuklinski (The Hague: Mouton, 1978), p. 170.

66. Stohr, "Some Hypotheses on the Role of Secondary Growth Centres," p. 85.

67. Howard Spodek, *Urban-Rural Integration in Regional Development: A Case Study of Saurashtra, India, 1800-1960,* Research Paper 171 (Chicago: Department of Geography, University of Chicago, 1976), p. 109.

68. Policies are outlined in D. C. Rao, "Economic Growth and Equity in the Republic of Korea," *World Development* 6, no. 3 (1978): 383-396; Gustav Ranis, "Equity with Growth in Taiwan: How 'Special' Is the 'Special Case'?" ibid., pp. 397-409.

69. Sen-Dou Chang, "The Changing System of Chinese Cities," *Annals of the Association of American Geographers* 66 (September 1976): 398-415.

70. Ibid., p. 411.

71. Ibid., p. 414.

72. Ibid., p. 415.

73. Jan S. Prybyla, "Changes in the Chinese Economy: An Interpretation," *Asian Survey* 19 (May 1979): 408-435; David Morawetz, "Walking on Two Legs? Reflections on a China Visit," *World Development* 7 (1979): 877-891; Suzanne Paine, "Balanced Development: Maoist Conception and Chinese Practice," *World Development* 4, no. 4 (1976): 277-304.

CHAPTER 6

PLANNING AND MANAGING
SECONDARY CITY DEVELOPMENT

The economic and political forces that have created inefficient and undesirable patterns of urbanization in many developing countries are unlikely to change automatically, and it is
equally unlikely that strong systems of secondary cities will
emerge without deliberate and consistent government inter-
ventions. But whether secondary cities grow spontaneously
or as the result of government policies, their development
creates new and more complex social, physical, and economic
problems. New demands are made by growing populations
for basic services and facilities, the extension of infrastructure and utilities, more and better jobs, housing, education,
and health care, and for the amenities usually associated with
urban living. Support services and infrastructure become
more important in the local economy, individual location
decisions create costs or benefits for the city as a whole, and
investment in social overhead capital becomes more crucial
for attracting investment in directly productive activities. The
ability of local and national governments to meet these
demands and satisfy these needs determines, to a large
extent, the capacity of secondary cities to continue to perform important development functions.

Thus national urbanization strategies must also anticipate
and seek to address the inevitable problems that accompany
the growth of secondary cities. Policies must focus on creat-

ing self-sustaining economies capable of meeting changing needs in secondary cities over the long run. The case histories reviewed here identify a number of common and recurring problems that emerge with greater population concentration and economic diversification, and provide strong clues about the kinds of programs needed in secondary urban centers. Underlying any strategy for secondary city development, however, must be a set of strong policies for increasing the income and purchasing power of the large numbers of poor people living in intermediate urban centers.

Reducing Poverty

Ultimately, governments in developing countries must come to grips with the growing poverty in secondary cities, not only because it constitutes a serious social problem, but because the poor make up a large percentage of the population that must be incorporated into the local labor force and among whom demand for consumer goods and services must be increased. As long as a large percentage of the population in secondary cities remains at or near subsistence income levels, little real progress can be made in strengthening their economies.

Although accurate data on the magnitude of poverty in secondary cities are sparse, scattered evidence indicates that in most from 30 to 60 percent of the population have incomes below the poverty line set by national governments or international agencies. In Brazil, from half the families in southern cities such as Curitiba to about 77 percent of those in northeastern cities such as Recife were living in poverty in the late 1960s and early 1970s. World Bank analysts concluded that "poverty is common in Brazil even in the apparently prosperous southern cities."[1] By the World Bank's standards of "absolute poverty"—per capita income equivalent to $50 or less per year and at least one-third below the national average per capita income—about 16 percent of the families in Recife and about 30 percent of the families in Natal were living in destitution. However, these analysts estimate that "lack of employment, inadequate urban services, squalid housing, malnutrition and disease characterize the living conditions of three-quarters of the population in cities like Recife and Natal."[2]

In Oaxaca, Mexico, about 60 percent of the households live at or below subsistence levels. Household income for poor families averages about $89 per month; about 65 percent of that income goes for food, with much of the rest taken up by shelter and other basic necessities. Thus little is available to save or invest in sideline activities that might increase the incomes of the poor or to purchase other goods and services.[3]

In the Philippines, about 53 percent of urban families are poor, and more than 40 percent are living in absolute poverty. Studies of conditions in Philippine secondary cities conclude that "economic opportunities fall far short of the needs of the poor."[4] Despite some stability in Davao City's economy during the 1970s, at least half the population survived on incomes equivalent to 28 cents a day at a level of absolute poverty well below that defined by the World Bank. Moreover, income distribution in Davao is highly skewed. The top 20 percent of the households receive 51 percent of the city's total income, while the bottom 20 percent of the households receive about 5 percent. Such a distribution provides little hope that cities such as Davao can strengthen and diversify their economies internally or generate the level of demand that will expand local businesses and industries. The people in the lower 60 percent of the income scale in Davao have only 28 percent of the city's purchasing power, while more than 70 percent of spendable income is concentrated in the hands of a small number of upper-income families.[5]

The present income distributions in most secondary cities of the developing world are not conducive to generating high rates of economic growth, and thus the conventional dichotomy between welfare programs and productive investments looses its meaning. Improving the living conditions of the poor, expanding their skills, raising their levels of educational attainment, improving their health, and raising their incomes must be viewed as long-range investments in the internal economies of secondary cities, not as welfare programs. Substantial evidence supports the argument that human resource development is the most effective means of alleviating poverty in developing countries and of increasing the productivity and incomes of the poor.[6] The "welfare program-productive investment" dichotomy is not only inaccurate, but misleading. Both indirect investments that support greater productivity through human resources development and direct investments in productive activities are needed to strengthen the economies of secondary cities, and one is not likely to have much impact without the other.

Four essential elements of a secondary city development strategy include: (1) increasing the quality and coverage of basic social and municipal services, facilities, and infrastructure; (2) improving physical infrastructure so that secondary cities can attract and support more diversified economic activities; (3) strengthening the economic base and employment structure to raise productivity and income and increase the capacity of secondary cities to continue to grow and diversify; and (4) building the administrative, planning, and financial capacity of secondary city governments to manage their development more effectively in the future (see Figure 6.1).

Extending Support Services and Facilities

The ability of governments to generate growth and development in secondary cities will depend in part on their capacity to meet the growing needs for basic social services, municipal facilities, and public infrastructure. Among the

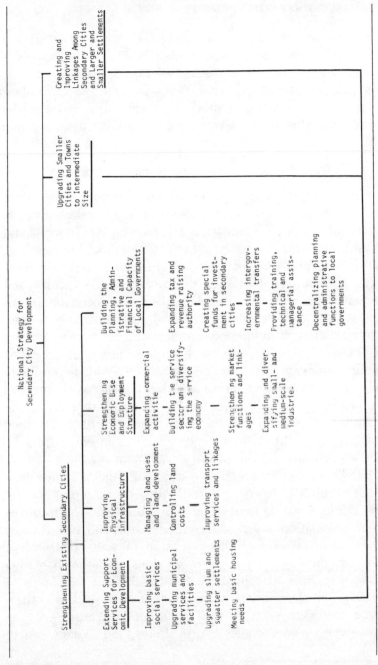

National Strategy for
Secondary City Development

Strengthening Existing Secondary Cities

Extending Support
Services for Econ-
omic Development

- Improving basic
  social services
- Upgrading municipal
  services and
  facilities
- Upgrading slum and
  squatter settlements
- Meeting basic housing
  needs

Improving
Physical
Infrastructure

- Managing land uses
  and land development
- Controlling land
  costs
- Improving transport
  services and linkages

Strengthening
Economic Base
and Employment
Structure

- Expanding commercial
  activities
- Building the service
  sector and diversify-
  ing the service
  economy
- Strengthening market
  functions and link-
  ages
- Expanding and diver-
  sifying small- and
  medium-scale
  industries

Building the
Planning, Admin-
istrative and
Financial Capacity
of Local Governments

- Expanding tax and
  revenue raising
  authority
- Creating special
  funds for invest-
  ment in secondary
  cities
- Increasing intergov-
  ernmental transfers
- Providing training,
  technical and
  managerial assis-
  tance
- Decentralizing planning
  and administrative
  functions to local
  governments

**Upgrading Smaller
Cities and Towns
to Intermediate
Size**

**Creating and
Improving
Linkages Among
Secondary Cities
and Larger and
Smaller Settlements**

Figure 6.1 A Framework for Secondary City Development Strategy: Micro-Policies

most important problems plaguing secondary cities are the growing deficiencies in basic social, health, and educational services, inadequate municipal sanitation and water services, and the paucity of power and sewer facilities. Secondary cities have extensive and spreading slum and squatter settlements, and they face increasing demands for basic housing and shelter.

IMPROVING BASIC SOCIAL AND MUNICIPAL SERVICES

Extension of social and municipal services has fallen far behind the pace of population growth in secondary cities in nearly all developing countries. The growing gap in social and physical services not only inhibits economic development, but acts adversely to maintain lower-income families in poverty. A number of studies indicate that the quality and coverage of services in secondary cities have been deteriorating over the past two decades. Local governments have been able neither to maintain existing services nor to extend them to those areas of the city with large numbers of migrants. Osborn observes the "general condition of decrepitude in Indonesian towns and cities, and lack of infrastructure and production of basic necessities." He argues that all of the secondary cities in Indonesia "need upgrading of streets, port and airport facilities, more and rehabilitated schools, hospitals, water service, drainage and government buildings."[7]

Public services in secondary cities of Thailand have been either expanding slowly or deteriorating over the past decade. The World Bank estimates that only about one-fourth of the urban population is served by piped water, and that the low levels of investment in the extension of water distribution systems, the high costs of connections, and the increasing costs of services all contribute to the relatively low levels of coverage in cities other than Bangkok. The lack of waterborne sewerage systems in Thai secondary cities and the prevalence of open canals, storm drains, and ditches lead to sanitation, health, and flooding problems. The quality and coverage of health and educational services in Thai secondary cities is much worse than in Bangkok, but usually better than in rural areas. World Bank analysts argue that the government's goals of achieving social equity and promoting more dispersed growth both require "reducing the disparities in publicly provided services and infrastructure between Bangkok and the provincial cities."[8]

Mabogunje describes problems in Nigerian secondary cities that can readily be seen in other African countries as well. He notes that with the lack of sanitation facilities and the inadequacy of sewage systems, "a notable feature of Nigerian cities is the concentrated flow of slop water from most houses to nearby lanes," where it is absorbed in the soil, atracts insect pests, and breeds mosquitoes. "Apart from the eyesore effect of such impeded water flow," he argues, this situation creates severe health hazards and obstructs the free movement of urban residents. "In addition, houschold sewage has to be head-loaded to disposal points or dealt with through other means such as the provision of pit latrines."[9] Weiker reports in his study of Eskisehir that "major problems like the lack of a sewer system (almost all sewage in Eskisehir is handled by septic tanks, which are fast becoming a serious health hazard), extremely inadequate public transportation and unsanitary conditions along much of the banks of the Porsak were all too familiar to long time readers of the local press."[10]

In nearly all regions of the developing world, the quality and coverage of services tend to be worst in the poorest neighborhoods. Studies of Philippine intermediate cities note that, almost universally, environmental sanitation is lacking in low-income districts. Pit latrines and open drains serve as sewers, and they usually contaminate ground water supplies. Garbage collection is inadequate and, indeed, the garbage that is collected from other parts of the city is often dumped in or near poor neighborhoods.[11] Basic health, educational, and social services are often less accessible to lower-income groups in secondary cities than are physical facilities and municipal services. Hackenberg summarizes the situation of lower-income young people in many secondary cities of the Philippines when he notes that in Davao the large percentage of them are "out of school, out of work, and out of luck."[12] Their lack of access to the basic educational, health, and training services that might improve their income earning potential and productivity traps them in a cycle of poverty. When poor families in Davao cannot obtain adequate income, they reduce household expenditure on the very services and facilities they need to make them more productive and to maintain a minimal standard of living—education, health care, and housing. Hackenberg correctly points out that these reductions of basic necessities are "counterproductive in the long run since they tend to divert people from production of consumer goods essential to raise living standards" and force them to accept "lower per capita income from the traditional sector." These survival mechanisms create increased squatting and idleness.[13]

Investments in services and facilities in secondary cities have been inadequate to keep pace with population growth and attract professionals, skilled workers, and private investors even in countries that have had relatively high rates of economic growth. World Bank studies conclude that urban services are inadequate in nearly all of Brazil's secondary cities, for example, and that the "north and northeast parts of Brazil are very poorly served, with the peripheries of the large cities being the worst off." Bank analysts found that more than 70 percent of the urban population remain unserved by sewerage facilities and that around 40 percent have no access to safe water, with the percentages much higher in some cities.[14]

Until the deficiencies in basic social and municipal services can be reduced, it is unlikely that secondary cities will be able to attract the productive investments that now go to the largest metropolitan centers, increase the productivity of their labor forces, or expand the purchasing power of their residents significantly. They must not only extend these services to larger numbers of people, but they must do so in ways that are cost effective, employment generating, and appropriate to local needs and conditions.

A good deal of evidence suggests, for example, that basic health care can be provided at lower cost in developing countries by converting from delivery systems that are exclusively or primarily curative to those that place greater emphasis on disease prevention. Greater emphasis must be placed on community and neighborhood clinics that have modest equipment, mobile service units, regular health surveillance within communities through the schools, programs for innoculation and vaccination, assistance in improving nutrition, family planning, and maternal and child care.[15] Many of these services can be provided by paraprofessionals rather than by physicians and nurses at much lower costs than through conventional health programs, while also creating new job opportunities for nonprofessional health workers.

Improvements in piped water systems, sewerage and drainage facilities, and waste disposal must be made simultaneously with the extension of preventive health care services, for inadequacies in these public services are a common cause of health problems in secondary cities. Improvements are required as well in the maintenance of existing systems, especially in older urban areas. Leak detection and repair programs are needed for urban water systems, as are surveillance activities to reduce waste and pilferage. Economic benefits can be obtained from the extension of water systems by helping small-scale manufacturers in secondary cities to produce construction equipment and water system components such as pipes, valves, faucets, and meters, rather than importing them from the capital city or from abroad. Temporary solutions may be necessary in low-income neighborhoods.

The most appropriate solutions to service problems may be in providing projects that assist residents to sink wells or to build small-scale treatment stations so that water can be drawn from untreated sources. Technical assistance can be offered to help low-income communities build small reservoirs and collect rainwater more efficiently.[16]

A combination of conventional and adaptive technology must also be used in extending sewerage services in secondary cities. Sanitary disposal of waste water is no longer an amenity in cities that reach 100,000 or more population; it is an essential part of an urban environmental health program. In addition to improvements in pit latrines, surface sewers, and treatment installations, improvements in garbage and night-soil collection are essential. In low-income neighborhoods on the periphery of secondary cities it may be possible to build community cesspools with tanker cleaning systems that distribute waste water to nearby agricultural areas or to stabilization ponds that can be used for fish or poultry raising or fodder production.[17]

Permanent solutions require extending trunk sewers, expanding treatment plants, and maintaining newly constructed facilities. The extension of these services can bring economic as well as social benefits to secondary cities when labor-intensive technology is used to construct and manage water and sewer systems and when components and services are procured from local establishments.[18] Moreover, increased attention must be given to providing and locating public services and facilities in ways that minimize energy costs, which have been rising steadily in oil-importing countries. Emphasis should be placed on providing public utilities that can use biomass conversion to fuel. Studies have shown that small and middle-sized cities suffer much more from increases in energy prices and declines in supply than the largest metropolitan areas because they have less flexibility to substitute energy sources.[19]

UPGRADING SLUMS AND SQUATTER SETTLEMENTS
AND MEETING BASIC HOUSING NEEDS

Opportunities also exist to provide basic housing and shelter in ways that promote economic development in secondary cities. Housing shortages are acute, and slum and squatter settlements abound in nearly all secondary cities in the developing world. Public housing construction simply cannot meet the demand for basic shelter, and few public or private financing organizations exist to provide low-income families with the capital needed to buy or build homes. Mabogunje notes that in Nigeria, government policies have had little impact on the worsening housing problem. Despite public investment programs, the amount required for an initial deposit and the monthly mortgage payments for houses have meant that "a large part of government action has resolved the housing problems of only a minority of the population in the higher income groups."[20] As a result, Nigeria's secondary cities are overcrowded, squatting has proliferated, and the conditions in slum neighborhoods have worsened.

Appropriate financial and administrative arrangements for dealing with housing problems simply do not exist for the vast majority of people in most developing countries. In secondary cities of the Philippines, for example, there is no local mortgage financing available for lower- and middle-income housing, and local governments have few or no resources for building or financing low-cost shelter. Only a few of the many families needing housing financing can qualify for public loans on concessional terms. Sites appropriate for community improvement programs in secondary cities are usually privately owned, and funds needed for improving services and facilities in low-income neighborhoods are scarce.[21]

Even in more prosperous developing countries it is difficult for the majority of people living in secondary cities to own their own homes or apartments. Dwelling prices increased by 27 percent a year between 1970 and 1977 in Korea, twice the annual rate of increase in the urban consumer price index. The annual increase in the cost of construction materials rose by 11 percent, while construction costs rose by nearly 19 percent a year. Dwelling prices in Korea increased 4 or 5 times more than household income. Because they must rely on household savings to purchase a home, the vast majority of people are eliminated from the housing market. It would take an average family living in Seoul or Pusan 37 years to save a sufficient amount to buy a house at 1979 prices, and it would take about 22 years for a family living in an intermediate city to save an adequate amount. Nearly 90 percent of those Korean urban families that were able to purchase dwellings in the 1970s had to rely on their own funds; less than 10 percent were able to borrow money, and most who did so received loans from other individuals.[22]

Conventional housing markets, urban renewal, or public housing simply cannot provide solutions to the slum and squatter problems of secondary cities. Strong arguments have been and should continue to be made against public projects that wipe out slums or drive out squatters prematurely. Although they may be visible evidence of poverty that is discomforting to local government officials and unpleasant reminders to the elite of the dual economic structure that exists in most secondary cities, slum and squatter settlements house large numbers of people for whom better alternatives often cannot be found. In addition to offering affordable housing to low- and middle-income families, slum and squatter settlements are often conveniently located near the bazaars and markets on which the poor depend for food, basic necessities, and employment. Many squatter areas contain large numbers of "shop-houses" from which the poor offer low-cost services or sell basic consumer goods, and the combination of shelter and employment in squatter districts reduces transportation costs.

Much more attention must be given to sites-and-services projects that allow poor families to improve their homes, add to and build upon basic dwelling units, and expand their structures slowly as their incomes increase. Sites-and-services projects can help low-income households through community improvement programs, the extension of water and sewerage facilities, electrification and street lighting, health clinics and preventive health services, and construction training programs. Housing finance policies should provide low-cost loans to families for upgrading basic dwellings and for purchasing the land on which they are built. Moreover, public housing programs must begin with simple core dwellings that can be expanded and improved incrementally.

Publicly constructed or financed housing for low- and middle-income groups can be a stimulant to local economic development, rather than a drain on local and national resources, if housing and infrastructure construction programs are planned as part of an overall economic development policy for secondary cities. Experiments with the construction of housing and public infrastructure as "leading sectors" in secondary cities in Colombia, for example, have generated demand for a wide variety of building materials, equipment, and durable goods and, through backward and forward linkages to related industries, can generate indirect employment as well as provide jobs directly. When plans are made for integrating housing construction through backward, forward, and lateral linkages with local building, material, and supply industries, for training local workers in construction skills, and for channeling increased income back into the purchase of locally produced goods and services, housing investments can have a significant impact on secondary city economies while providing needed shelter for middle- and lower-income families.[23]

Improving Physical Infrastructure

As secondary cities grow, they face more complex and serious physical problems. Planning effectively for their physical development, however, is often constrained by the ways in which their physical growth occurred—usually by accretion rather than through transformation. The ecological and physical structures of Third World cities were shaped by many forces and, as noted earlier, evolved with traditional and modern activities, urban and rural functions, and new and older districts expanding side by side. Abu-Lughod points out that cities in the Middle East contain three or more subcities: the historic precolonial core, the modern colonial appendage, and the unregulated indigenous quarters on the outskirts.[24] But she notes that close examination of these cities reveal that they may contain up to six types of subcities, including the medina core, modern appendages built in European style during the nineteenth and twentieth centuries, transitional working-class zones that grew between the medina core and the modern appendages, rapidly proliferating and uncontrolled settlements within the city built by rural migrants, suburbs containing the homes of upper-class and high-income families, and rural fringe developments that have engulfed rural villages.[25]

The physical development of many Asian secondary cities also came about through accretion of new functions during different periods of colonial rule. Barringer notes that the ecological structure of Taegu, Korea, during the late 1960s and early 1970s was a "mixture of traditional functions (e.g., the herb market and farm-oriented markets), Japanese urban planning (functional for the Japanese during their occupation), contemporary *ad hoc* industrial and migratory growth, with an overlay of national-level planning (the new Taegu railway station and associated highways)."[26] Studies of Ahmedabad, India, found that it

> developed through the addition of land and through population growth in relatively distinct historical stages. The old city, within the fortress walls, dates from the fifteenth century. The pols and old pura organization on the west bank are associated with the

mid-eighteenth century. The industrial section just outside the walls, with its workers' quarters and squatter settlements, was established in the nineteenth century. The development of residential areas in the far-eastern and western sections followed the economic restoration of the city. The establishment of administrative and commercial centers on the west bank is even more recent.[27]

In cities that expanded through the addition of new functions and the segregation of social and economic activities in different districts, some areas of the city took on distinct social, economic, and physical characteristics, while others acquired a mixture of those found in other parts of the city. Thus it is often difficult, if not impossible, to prescribe physical controls for or to plan for the physical development of the city as if it were a homogeneous entity. Each subarea of the city must be analyzed and planned individually. Three problems that are common to nearly all secondary cities, however, are rapidly rising land costs, unregulated and inefficient patterns of land use, and increasing demands on intracity and interurban transportation systems.

MANAGING LAND USES AND CONTROLLING LAND VALUES

Nearly all of the case histories of secondary cities reviewed here note that population growth and economic diversification brought rising land costs within the cities and on their fringes. United Nations studies estimate that land prices in larger cities of developing countries have risen by 10 to 20 percent more than consumer price indexes, thus making it more difficult for lower- and middle-income families to obtain urban land for house building, thereby increasing the number of squatters.[28] In Korea, for example, residential land prices in urban areas have increased by 34 percent a year, a rate of increase, nearly 3 times higher than that of the urban consumer price index.[29] The burdens of rising land costs often fall most heavily on low-income families. In the

Philippines, land in most secondary cities is scarce and priced beyond the means of low- and middle-income families. The poor must rent the land on which they build their homes, and often landlords maximize income by renting to as many families as can be squeezed onto a plot. Usually few if any services are provided by landlords and the overcrowding is made worse by the lack of water and sewerage facilities.

Weiker also notes in his study of Kayseri and Eskisehir, Turkey, that although a good deal of capital investment was made in housing, infrastructure, and public buildings, little of it appeared to benefit lower-income groups. "Rather, current property owners were reported to be receiving high prices—and land values have risen enormously in Kayseri, reportedly to one of the highest levels in Turkey—after which the improved land is resold at prices which do not cater much to those most in need of improved facilities."[30]

The costs of land tend to rise so rapidly in secondary cities because, in most, land speculation is the best hedge against inflation, monetary devaluation, and economic fluctuations for the wealthy. Measures can be taken to "disinflate" the cost of land through progressive taxation on unused land, higher taxes on owners of more than one home, and government intervention in the land market, but these are unlikely to be effective until local and national economies offer better and more diverse opportunities for investment.[31] But some action is needed because the rising costs of land increase the costs of doing business in secondary cities, push people from villages on the urban fringe, accelerate the conversion of agricultural land to other uses, promote sprawl, and increase the costs of acquiring land and rights-of-way for public utilities, services, and roads. In many secondary cities rapid increases in land costs can further concentrate land ownership under the control of a few wealthy families or large corporations and generate more widespread land speculation.

Moreover, many case studies note the increasing difficulties that residents of secondary cities face because of the lack of control over land uses, the spontaneous development of the city, and the location of residential, business, and public activities without regard to external costs. Congestion and rising costs of transportation are often cited as the most immediate and visible signs of uncontrolled physical development, but they are usually followed by other problems that are less visible and more costly. Pannell found that the inability of local government in Taichung, Taiwan, to control physical development led to discontinuous and segregated land uses on the city's fringe—especially in and around areas accommodating large industries—that led to increased traffic congestion and rising costs of transportation for people working in those industries.[32] In Chiangmai, Thailand, construction of large facilities such as a teachers' college, a bus terminal, and private commercial buildings with little or no land use planning stimulated the expansion or location of other activities near them over which the city government could exercise little or no control. Observers note that public agencies and private investors alike could "construct virtually any kind of building they prefer."[33] Municipal regulations were concerned only with structural soundness and fire safety, and gave officials little or no control over land use relationships or ability to guide land use to minimize the costs of energy, transportation, or public facilities.

Ironically, secondary city governments often adopt building codes and land use standards from Western industrial countries that are not only inappropriate but that may create unnecessary problems. In Nigeria, for example, overly restrictive housing construction standards constrain housing production and increase housing costs. By lowering density controls, lot coverage, room floor area requirements, and height controls, Nigerian officials could increase housing production, make units less expensive, make land use more efficient, lower the costs of public services, and restrain sprawl.[34]

As in many other aspects of urban development, secondary cities are often caught in a vicious cycle of land use problems.

Weiker notes that "the physical problems of Turkish cities are often so overwhelming and the cities' financial resources are so grossly inadequate that there is simply no time or energy for matters other than the immediate physical problems."[35] But in secondary cities of Turkey, Brazil, and Malaysia, Rivkin found that local and national governments had few if any effective means of solving even immediate problems. He found that in these countries, as in many others, the inability to control or guide physical development added to the rising costs of land and made it more difficult for local governments to find adequate and appropriate sites for low-income housing, schools, health services and other public utilities, services, and infrastructure. The worsening air and water pollution and the degradation of environmental quality in many secondary cities is largely attributable to inadequate or nonexistent land use controls.[36]

Rivkin and others have identified actions that national and local governments can take to control land costs and to guide land use in secondary cities. They emphasize the urgency of taking action before cities grow so large and complex that locational decisions are irreversible and uncontrollable. The actions include:

(1) infrastructure provision—the location of services and facilities by government in such a way as to encourage or discourage those activities dependent on them in various parts of the city;

(2) land acquisition in advance of need—government purchase of land to guarantee its availability for construction of public facilities, parks, office buildings, or recreational areas or to preserve open space needed to maintain environmental quality standards;

(3) land banking—acquisition and reservation of land for later use or to control the density of development in various parts of the city;

(4) government construction and financing—to ensure the provision of housing, industrial, or commercial facilities in appropriate locations in the city;

(5) public-private development instruments—acquisition of land by government and reparceling to private developers for controlled uses;

(6) urban renewal—removal of dilapidated buildings by government, acquisition of land, and redevelopment in new or improved uses by government or private developers;

(7) land use regulation and controls—zoning, subdivision, and building ordinances that guide or restrict private use of land;

(8) building permission requirements—government controls or regulations over building types and locations, land coverage, and structural standards;

(9) value freezing—government authority to freeze site values of land designated for future construction of public facilities or infrastructure to discourage speculation; and

(10) taxation—special levies on land held out of development for speculative purposes or on windfall profits from land values rising because of public betterment on adjacent or nearby land.[37]

Some combination of appropriate controls with relevant land use planning can assist local governments to guide physical development of secondary cities before they grow in ways that are economically and socially detrimental to further expansion.

IMPROVING TRANSPORTATION SERVICES AND LINKAGES

A closely related problem is the severe strain placed on transportation services, roads, and highways as secondary cities grow in size and density. Policies are needed to improve intracity transportation services and to strengthen intercity transport linkages.

The World Bank reports that public transport networks have not been extended in relation to patterns of urban growth in most developing countries and that severe deficiencies in services, especially in squatter and low-income neighborhoods, increase the living costs of the poor as well as the transportation costs of businesses and industries. The inadequacy of public transportation systems is exacerbated by the increasing use of automobiles in larger secondary cities that were not designed for the efficient flow of automobile traffic and that cannot be redeveloped easily to accommodate increased automobile use. Mabogunje notes that in Nigerian cities "the transport situation, especially in major metropolitan centers such as Lagos, Ibadan, Kaduna and Kano, is a great cause for alarm." He notes that the unrestricted importation of cars has created a situation in which "most Nigerian cities now find their ratio of roads per square mile of buildings inadequate to insure a smooth flow of traffic. Many of the cities have no more than two or three arterial roads."[38] In most secondary cities of Asia and Latin America, especially in their early stages of economic growth, both public transportation services and the road systems were inadequate to handle increased movement and traffic.

World Bank analysts suggest that plans for improving transportation and road systems should consider technology that is appropriate to the current size and social and economic characteristics of secondary cities, rather than technology used in the large metropolitan centers of Western industrial countries. Cities with only about 100,000 in population may not need extensive or costly investments; because they are relatively small, only short trips are required to reach most social, economic, and recreational activities.

In many cities mixed land use patterns provide people with easy access to nearly all activities. Most people depend on public transportation services for longer trips, and on bicycles, motorcycles, cabs, minibuses, or nonmotorized forms of transportation for medium-length trips. Special attention should be given in these cities to properly managing through traffic to prevent peak-hour traffic jams, to low-cost forms of nonmotorized transportation, to providing adequate thoroughfares for pedestrians and cyclists, and to maintaining efficient forms of public transportation, especially flexible minibus services. The transport needs of these cities can often be handled by small-scale enterprises—by pedicab, taxicab, minibus, and even trucking services—or through the informal sector.[39] More effective land use planning is needed to minimize travel needs and transportation costs.

As cities grow to from 250,000 to 500,000 in population, transportation problems become more complicated. As densities increase and the city physically expands, the volume of movement and average trip lengths increase substantially, requiring more efficient public transportation systems and a wider variety of transport modes. Traffic management problems increase and a larger number of potential bottlenecks and centers of congestion emerge, especially at transport terminals, markets, business centers, and other places that attract large crowds of people. Increasing dependence on motor vehicles tends to congest those parts of the city with narrow streets, where the physical pattern of development is not conducive to road widening or massive redevelopment. In these cities, greater reliance must be placed on public transportation systems, especially bus and minibus service, and on controlling land uses and construction in ways that minimize travel requirements. More and better roads must be built, traffic must be managed more efficiently, and, if the city is a transfer point in an intercity system, the location of terminals and highway connections must be planned carefully so that intercity traffic does not exacerbate the problems created by higher volumes of intracity traffic. The importance of planning ahead when cities begin growing for the kinds of traffic and transportation problems that metropolitan centers face has been emphasized by World Bank studies, for many of the cities of from 100,000 to 500,000, "particularly in Latin America, Africa and East Asia, are growing rapidly and will in a decade or less fall into the next larger size grouping, where urban transport becomes a major problem for city management unless steps have been taken early on ensuring an appropriate urban transport pattern."[40] Bank analysts argue that the "most important aspect here is the encouragement of a land use pattern which reduces the need for lengthy commuting trips by ensuring a balanced spread of employment opportunities as the city grows."[41]

Intercity transportation linkages can be even more important to the economic development of secondary cities. Intercity highways, railroads, and other forms of transportation linkages play an important role in providing access for rural people to the services and facilities located in secondary cities, in increasing distributional efficiency, and in lowering the transport costs of businesses that depend on intercity trade. Most of the larger industries in Thailand's regional urban centers are located along major highways, indicating the importance of access in locational decisions.[42]

The extension of interurban highway or rail linkages has a number of effects on the cities and towns that are connected. It can increase interurban trade, yield production economies, and lower per unit transport costs. It can generate increased traffic and thereby further decrease shipping costs, improve services, and induce industries to locate along the transport axis at secondary nodes or on the periphery of intermediate cities. World Bank analysts argue that "the development of economic activities along the major transportation corridors should be encouraged, not discouraged. The possibility of the spread of manufacturing activities will also be related to the opening of new transportation links connected to the main corridors."[43]

However, it should also be noted that while the extension of major transportation linkages can have a positive impact on articulating and integrating the settlement system in developing countries, it can generate massive migration from rural areas and smaller towns. Transportation and road improvements increase rural people's access to commercial and personal services, job opportunities, and public facilities in secondary cities, but they also reduce the costs of migrating from rural areas and can therefore stimulate the flow of migrants at a more rapid pace than secondary cities can absorb them.[44] Rivkin has pointed out that the construction of Turkey's highway network, which increased access to interior towns during the 1950s and 1960s, stimulated the flow of rural migrants to urban centers. At the same time that the highway network was being extended, the number of urban centers with more than 50,000 residents increased from 12 to 27, and their share of urban population increased to more than 50 percent. All 27 cities were linked to the highway network. "In this case, it is clear that the massive public investment in highways made at least a major contribution to the distribution of urbanization across the country," Rivkin concludes, and the network helped to promote the growth of a widely dispersed system of secondary urban centers in interior regions.[45] The positive impacts from increasing intercity transportation linkages thus may only appear if the economic base and employment structure of secondary cities along the routes or at the nodes can be expanded and strengthened.

Strengthening the Economic Base
and Employment Structure

A third set of problems that national urbanization strategies must address is that of strengthening the economic bases of secondary cities to generate more employment opportunities. This requires strengthening their commercial and market functions, especially the informal sector and small-scale enterprises, and increasing the productivity and diversity of small- and medium-scale industries. Expanding and diversifying the secondary and tertiary sectors may be the most direct and effective way of generating the employment that will increase the income of the poor, expanding internal demand for locally produced goods and services, and increasing the capacity of secondary cities to absorb larger populations. Those industries and businesses that are most likely to absorb the unemployed should receive the greatest attention in assistance programs.

Studies of unemployment in secondary cities of Colombia show a pattern characteristic in other countries as well. Most of the unemployed are young, about two-thirds are under 25 years old, and about 80 percent are under 35 years old. Many—about 40 percent in Colombia—are new entrants to the work force, having little or no experience. Existing members of the labor force who are unemployed tend to have low levels of education and training and some are both illiterate and unskilled. Moreover, many of the unemployed are local residents who were born in or around the city, although migrants tend to make up a substantial percentage of the new entrants to the labor force. Not surprisingly, many of the unemployed are poor or come from low-income families. Because of their young age, most are dependents rather than family heads, although the burdens of unemployment tend to be greater for those who do have families. Most of the unemployed seek jobs in the service sector or jobs as artisans, laborers, semiskilled workers, or office or sales workers.[46]

Programs that expand and strengthen the tertiary and small-industry sectors would probably be most suitable for reducing the high levels of unemployment and underemployment in secondary cities, whereas large-scale industries would probably only have indirect and uncertain "trickle-down" effects. Large industries are likely to provide the best opportunities to skilled workers who already have jobs, and would most likely be export oriented rather than tied to local demand. In most secondary cities priority should be given, at least in the initial stages of development, to expanding the marketing and commercial functions that already exist and to strengthening the cities' roles as trade and service centers.

STRENGTHENING THE COMMERCIAL, SERVICE, AND MARKETING BASE OF SECONDARY CITY ECONOMIES

Hackenberg quite correctly points out in his study of Davao City that although the tertiary sector provides the largest amount of employment in most cities in the Philippines, "the role of the bazaar economies as a mobility multiplier capable of accelerating the transition from low- to middle-income status is precarious."[47] He notes that the demand elasticities for subsistence goods—food, clothing, and household items—sold by measure, for negotiated prices, and by large numbers of vendors and traders, are very limited and that most market stall owners or street vendors simply do not have the resources to generate greater demand for their goods. Problems in the informal tertiary and small-scale commercial sectors are exacerbated by rural migration because the tertiary sector in most secondary cities cannot absorb new entrants without diminishing the share of earnings for those already in it. New entrants to the informal sector often further constrain upward mobility for the poor.[48] Bromley arrives at similar conclusions in his study of Cali, Colombia. He notes that a substantial majority of those engaged in informal sector activities in the city derive only subsistence or less than subsistence income and that the great majority of them "have to face up to continuing poverty, great economic and occupational insecurity, frequent problems with authorities and ongoing dependence on suppliers of merchandise, capital and equipment."[49]

In many secondary cities with less than 300,000 people, markets act as the major sources of bazaar, informal, and small-scale commercial employment, and play crucial roles in linking secondary cities with their rural hinterlands. Thus policies aimed at strengthening their effectiveness in regional trade can have an important effect on expanding secondary city economies. But care must be taken in choosing viable markets for development. Studies of Philippine markets suggest that their volume of sales, scale, and distance from larger markets are critical factors in determining profitability. The smaller town and city markets that were studied were not profitable public enterprises because their operating costs were high, their pricing policies were constrained, and their small size limited their attractiveness to people in the rural hinterlands. Analysts found that "there appear to be economies of scale in the operation of public markets. Our very limited and very crude analysis has indicated that these efficiencies become effective when markets service a population of around 40,000." The study found that although populations of that size in a municipality do not guarantee financial success, "the technology of market operations is a barrier against the profitability of small markets."[50]

A number of government policies might be used to strengthen the small-scale commercial and service sector in secondary cities, including direct government procurement of goods and services from small establishments and businesses, providing tax or other incentives for large-scale firms to subcontract with small-scale establishments, providing low-interest loans to small businessmen, and organizing marketing and purchasing cooperatives for them.[51] In some countries, government and institutional purchasing of goods and services from small-scale businesses can be the most direct and immediate way of strengthening them and expanding their employment potential. Observing the limited capacity of small-scale enterprises to expand the demand for their goods, primarily because of widespread poverty and low income of potential consumers in secondary urban centers in the Philippines, Hackenberg argues that a wide range of goods and services—printing, food, uniforms, maintenance equipment, furniture, vehicles, sign painting, and other services, for example—should be purchased by national, provincial, and local governments from small-scale producers.[52] In the short run, the costs of governments might be higher than if they were to buy in volume from large firms, but the long-run returns in the form of broader participation in the economy, a wider distribution of profits, and increased purchasing power among lower-income families who would benefit from expanded employment opportunities could well outweigh the added immediate costs.

Action must also be taken to assist migrants and the unemployed to acquire skills and seek jobs that will increase their income, purchasing power, and upward mobility, including: (a) training in relatively simple manual skills such as sewing, candy making, and masonry; (b) training for relatively skilled positions in the organized sector, including large-scale businesses and service enterprises such as plumbing, electrical work, welding, fitting, and carpentry; (c) training in basic management skills for small enterprises, such as rudimentary market product design, product packaging, purchasing, marketing, accounting, and credit financing; (d) assistance in finding employment for those who upgrade their skills through training; (e) assistance in setting up small businesses as individual enterprises and cooperatives; and (f) follow-up technical assistance to people who go into business after their training programs, refresher, or remedial training and supplementary training in more advanced skills.[53]

Ultimately, small-scale commerce and services in secondary cities can only be strengthened if internal demand and purchasing power are increased, especially among those lower-income families that currently do not have the resources to participate effectively in the urban economy.

STRENGTHENING AND EXPANDING SMALL-SCALE INDUSTRIES

It was noted earlier that small industries dominate the manufacturing sectors in secondary cities and that they are both a strength and a weakness of their economies. Small- and medium-scale manufacturing probably provides the best hope for expanding and diversifying the economic base of secondary cities in the short run, yet the very characteristics of small-scale industries tend to minimize their impact on production and employment. Small-scale industrialists face myriad problems in establishing themselves and surviving in developing nations. They often lack the skills necessary to identify good potential investments, to prepare proposals for external funding, to test the feasibility of potential investments, or to negotiate loans from commercial banks and government agencies. They are generally excluded from government incentive schemes that benefit large industries and lack access to human resources and markets needed to produce and sell their goods. They finance their activities either from family savings or from credit obtained at high interest rates from moneylenders, buyers, or suppliers. Small industries generally have low levels of productivity, poor quality output, inadequate technology, obsolete equipment, poor packaging, and limited and uncertain markets. In addition to their limited access to credit and finance, they generally have insufficient raw materials and lack the managerial skills and knowledge of modern marketing, production, and accounting methods that might help them to increase profits. Without resources, small-scale industrialists lack access to technical assistance and managerial advice. They are often discriminated against in government programs offering financial aid, subsidies, preferential purchasing, and export assistance to large industries. Small-scale industrialists often fail to survive and grow because they lack the knowledge needed to manage their enterprises efficiently.[54]

Hackenberg notes in his study of Davao City that industry accounted for a relatively small amount of total employment, and that small-scale industries, which dominated the manufacturing sector, provided only low-wage jobs. The median annual income for manufacturing workers in Davao was among the lowest for all occupations, except for manual and farm labor.[55] Roberts came to similar conclusions in his study of Huancayo, Peru. Manufacturing in that city has traditionally employed less than 20 percent of the labor force and the small scale of enterprises in the city is not conducive to increasing employment opportunities, returning only marginal profits and income to those who own them. Most of those engaged in small-scale enterprise earn only subsistence incomes.[56]

However, small industries do engage the labor of substantial numbers of people in developing countries and can, with proper incentives, provide a stronger base for economic growth and labor absorption. As noted earlier, countries that have made deliberate and serious efforts to expand small-scale industries have increased the number, distribution, productivity, and employment capacity of enterprises employing less than 50 workers. Korea, Taiwan, India, China, and, most notably, Japan in its early stages of economic growth provided the supporting services and investment incentives needed to expand small industries in secondary cities and in rural towns and villages.

A number of actions must be taken by governments to stimulate small-scale manufacturing in secondary cities. These include assisting small-scale industrialists with identifying investment opportunities, providing technical assistance to small-scale entrepreneurs in organizing businesses, helping them to obtain adequate supporting services, infrastructure, and facilities, increasing small-scale manufacturers' access to financial resources and credit, providing operating assistance and training in management and production, and helping them to expand demand and overcome the limitations of small size.[57]

Assisting small-scale industrialists with identifying investment opportunities. Small-scale entrepreneurs in secondary cities of most developing countries rarely have the resources to identify potentially sound investments. If small-scale manufacturing is to be expanded, government agencies, commercial banks, or development organizations that have the resources and professional competence to do so must assist entrepreneurs in uncovering appropriate and viable opportunities. Entrepreneurship in secondary cities is sporadic and the success of new ventures is often uncertain. Individuals or families commit small amounts of savings or cash obtained from moneylenders at high interest rates to small production activities for which they often have little information about the market or potential demand for their product, and may quickly fail unless they can get technical assistance in organizing and operating their businesses.

Low-income families are often unwilling to take risks or to face the uncertainties of starting new businesses. In many Asian and African secondary cities, the highest levels of entrepreneurship are found among minority groups or outsiders who are not bound by traditional obligations, practices, or perspectives. Without organized assistance or external stimulation, many potential opportunities in secondary cities can go unidentified and unexploited. But there are potentially large numbers of opportunities for small-scale enterprise in secondary cities. In regions where agricultural production and income have risen above subsistence levels, there is usually latent demand for goods and services currently imported from other regions or larger cities. This is especially true for local market goods such as furniture, clothing, housing construction materials, hardware, millwork, and fabrics. Opportunities exist for processing farm commodities and extracting natural resources, charcoal and brick making, light transportation and construction, and a wide variety of commercial services such as appliance, vehicle, and machinery repair, spare parts manufacturing, and agricultural equipment production and repair.[58] Agriculturally linked industries can also be established in subsistence farming areas where new seed varieties, irrigation, or farm-to-market roads are being introduced.[59]

Experience has shown, however, that to increase participation in small-scale manufacturing there must be public assistance programs that help entrepreneurs to identify opportunities and that create the conditions that allow new ventures to operate efficiently. India's experiments with pilot industries projects in regions where agricultural conditions are favorable and production is increasing, in cities with high levels of unemployment, in tribal and economically lagging regions, and in cities where there are large industries, branches of universities, or industrial and agricultural research institutes have shown that small-scale manufacturing opportunities can be expanded. Some countries have established regional technical assistance and service centers in secondary cities that help entrepreneurs identify potential investment opportunities. They conduct areawide industrial, economic, or market surveys that provide information to entrepreneurs and help them prepare finance or credit proposals. In some places they conduct entrepreneurial search, selection, and training programs. In cities where individual entrepreneurs may be unable or unwilling to take the risks of setting up their own small industries, these centers establish manufacturing cooperatives that are later taken over by individual owners or cooperative members.[60]

Helping small-scale entrepreneurs to organize new businesses. The low levels of education and literacy among many potential entrepreneurs in secondary cities make it difficult for them to prepare proposals for financing or credit that are acceptable to commercial lending institutions. These entrepreneurs' inability to hire consultants and management experts to prepare feasibility analyses and their lack of access to information concerning domestic and export markets often exclude their proposals from serious consideration. They are often at a disadvantage in choosing locations for their businesses; they generally lack the ability to assess alternative sites in terms of market characteristics, distance from major market centers, and the availability of transport and communications, raw materials and semifinished inputs, water and power, and skilled labor. Small-scale entrepreneurs need a great deal of assistance in formulating acceptable prospectuses or proposals for funding by commercial banks or government development corporations.

Some government agencies in Asia offer training courses that teach skills needed by small-scale entrepreneurs to formulate project proposals and analyze their feasibility. The courses cover the basic principles of marketing analysis, demand and supply forecasting, analysis of technology needs, plant and facilities location and layout, basic accounting and financial analysis, and elements of organizational and managerial feasibility assessment. The training is supplemented by field visits to successful plants and factories.[61] Similar courses adapted specifically to the needs of small-scale industrialists and conducted in secondary cities are needed to reach larger numbers of potential entrepreneurs.

Many small-scale manufacturers know little about appropriate production technology. They imitate other small operations, improvise machinery or buy discarded equipment, or simply use manual production methods when appropriate small-scale machinery could expand their output and increase their profits. In most countries, smaller industries are inefficient because equipment is inadequate or inappropriate to their needs, and plant layout and production design is faulty. Small enterprises often lack the capital to acquire the kinds of equipment that could expand their volume and quality of production and generate greater employment opportunities. And adequate physical infrastructure and utilities—roads, water, power, and waste disposal—are rarely available in secondary cities to support efficient operations. Plant and facilities for very small industries are often located in makeshift quarters or attached to the owners' houses, making efficient layout and design difficult.

The ability of small entrepreneurs to plan the production and marketing aspects of their projects efficiently is hampered, moreover, by their dependence on distributors or large buyers to set terms and prices and by unreliable contracting arrangements. Finally, family-owned enterprises are often too small to obtain economies of scale in production, access to reliable markets, or adequate supplies of raw materials—all of which complicate attempts to start new businesses. Many of these problems can be overcome through training and technical assistance programs.

Providing support services, appropriate infrastructure, and assistance with obtaining adequate sites. In Korea and a few other developing countries, governments provide assistance to small-scale industrialists in finding appropriate locations for their activities and offer leasing or hire-purchase options to make factories available to those capable of producing exportable goods but do not have the capital to purchase facilities.[62]

The inability of most small- and medium-scale producers to find adequate facilities has made construction of industrial estates an essential part of promotion programs in most developing countries. Generally, the government acquires and prepares suitable sites for industrial location, provides basic infrastructure such as roads, power, water, waste disposal, and service facilities, builds and modifies plants, and, in some cases, provides warehousing, common service facilities, repair shops, and financial services at the site. The World Bank suggests, however, that providing adequate infrastructure directly to secondary cities may be a more effective strategy than creating industrial estates. Studies conducted in Thailand conclude that

> although industrial estates are useful as a means for providing industrial infrastructure, their role as vehicles for industrial development is limited. Often the availability of industrial sites is not the critical constraint to industrial expansion in a city, and industrial estates can only accommodate a part of the new activities. Furthermore, some enterprises, especially small industries, may prefer private sites which may have more modest services but are cheaper than government provided sites. Finally, an industrial estate is as dependent as private development on basic infrastructure such as power supply and transport.[63]

Other countries provide assistance in relocating firms in larger cities to secondary and smaller urban centers. In Thailand, for instance, the Board of Industry can exempt a corporation from half the corporate income tax for a specified period of time, allow deductions from taxes for infrastructure construction and installation, allow double the normal deductions from taxes for increased costs of transport, electricity, water supply, and other utilities, and provide exemptions from business taxes on sales and import duties for imported materials used in production for firms that relocate from Bangkok to secondary cities or rural areas.[64] But the experience in Thailand and other countries shows that these incentives alone are inadequate to make the operation of medium-scale and larger industries efficient in secondary cities if basic investments have not been made in infrastructure, utilities, and transportation. The World Bank's report on the Philippines summarizes concisely the case for investment in these essential support services and facilities when it notes that "fiscal incentives without these provisions are unlikely to stimulate much new investment in the outer provinces and with such infrastructure, incentives are probably not needed."[65]

Increasing small-scale manufacturers' access to financial resources and credit. Perhaps the most serious problems reported by small-scale industrialists in developing countries are the lack of access to credit from commercial lending institutions, shortages of capital, and high rates of interest. Small-scale industrialists are considered poor credit risks. With small amounts of personal savings, meager assets to offer as collateral, and weak backgrounds in business, small-scale entrepreneurs are often excluded from serious consideration for loans by organized credit institutions, and most depend on family savings, small profits from other businesses, or high-interest, short-term loans from moneylenders. Once they go into debt, they are often caught in a nearly inescapable cycle of borrowing, and sometimes are exploited by lenders who impound their inventories or machinery as security.

The poor credit ratings of small-scale industrialists are often compounded by their lack of accounting and financial skills, their limited knowledge of banking and credit procedures, and the complexity of the requirements imposed by lending institutions. They easily become frustrated with complicated loan application procedures. In most countries, moreover, commercial banks and government agencies prefer to lend to larger, more established industries that offer less risk, greater security, and fewer problems. Small projects therefore are often appraised entirely on the basis of the credit worthiness of the borrower rather than on the merits and potential profitability of the proposed venture. Moreover, the cultural gaps between bank loan officers and small-scale businessmen, especially those from smaller cities, make the appraisal and negotiation of loans even more difficult.

All of these factors combine to restrict severely the access of small enterprises to capital and credit. But, if participation in productive activity is to be expanded in secondary cities, new means must be found to assist entrepreneurs financially. They must be trained in accounting, financial analysis, borrowing procedures, and loan negotiation. A wide variety of policies and programs must be established to increase their access to capital. Loan guarantee funds must be established or strengthened to protect the assets of government and commercial banks lending to small-scale industrialists, the amount of credit available to the small-scale industrial sector must be expanded drastically, and lending institutions must expand the number of low-interest, medium- and long-term loans on a concessionary basis. In most countries it is necessary for national development banks to provide refinancing and rediscounting facilities, and special concessionary loan departments must be established within commercial banks. The skills of small-industry loan officers must be improved so that they can appropriately appraise proposals submitted by small-scale entrepreneurs.[66]

At the same time, "integrated lending packages" should be created by government agencies and commercial lenders for specific industries to combine credit, technical assistance, design and marketing advice, and managerial training for small-scale borrowers. Lending procedures also must be modified and liberalized; more simplified appraisal criteria, flexible loan conditions, streamlined application and processing procedures, lower collateral requirements, provision of credit insurance at nominal cost, and single credit screenings on the basis of the proposal's merit rather than the credit rating of the borrower would all be essential to such a program.[67]

Much of Japan's success in generating small-scale industrialization can be attributed to the fact that an "impressive array of private and government financial institutions are engaged in financing small industries." As one observer notes, "No other Asian country has such a vast network of lending institutions to help small industries."[68] Where small-scale industrialization programs have been even moderately successful, as in Japan, Korea, Malaysia, and India, it has generally been due to the attitude among government and private bankers that this sector represents vast opportunities for expanding productive activity, income, savings, and investment, and that extensive efforts should be made to extend credit more widely to small-scale entrepreneurs and industrialists.

Providing operating assistance and training in management and production. It is only after small- and medium-scale industries are operating that many of the most severe problems become apparent to their owners. In order for such ventures to survive and expand, small manufacturers need technical assistance with operations and training in management and production. Internal management and control practices are notoriously poor among small-scale industrialists. Inadequate accounting and record-keeping systems, informal procurement and inventory methods, and lack of production

and distribution scheduling further contribute to their ineffi-
ciency. Training for workers and supervisors and "trouble-
shooting" services for small-scale industries are needed;
"common-service centers," containing facilities and equipment too expensive for individual companies to buy, should
be opened in secondary cities to provide access for all small
enterprises at a nominal fee.

*Helping to expand demand and overcome the limitations
of small size.* In some cases, establishments are simply too
small to be viable and advice should be available on the
possibility or desirability of establishing or joining industrial
production and marketing cooperatives, the opportunities for
becoming ancillary industries, establishing subcontracting
relationships with larger industries, or simply merging with
medium-size or larger establishments.[69] But even viable small-
scale establishments face difficulties because of their size.
Limited space for plant and equipment often makes small-
scale manufacturing or agricultural processing inefficient. The
difficulty of obtaining raw materials or their high prices
drives up costs, keeps machinery, equipment, and human
resources underutilized, and lowers profit margins. Small
enterprises generally cannot afford to stockpile large inventories of raw materials or finished goods, and products cannot be stored for sale when market prices are most advantageous. The nature of the product may allow it to be
produced only seasonally or when sufficient orders accumu-
late. The complexities of gaining access to export markets,
and their limited knowledge of domestic markets, keep small-
scale industrialists producing relatively small volumes. When
machines break down, the scarcity of replacement parts
either disrupts operations for a long time or workers improvise repairs, which, if done improperly, causes further damage
to machinery. The difficulty of hiring workers in some developing countries is due to minimum-wage requirements that
set salaries beyond levels that small, marginal-profit industries
can afford to pay. Exemptions from those laws for small-

scale enterprises may be needed to generate employment. Lack of trust and excessive fear of competition among some small-scale industrialists make them hesitant to seek external assistance or to delegate responsibility within their own establishments; they often attempt to do too many things themselves and thus disperse their energy and entrepreneurial talent. Technical, managerial, and supervisory training in these areas should be given as a part of every government or commercial bank loan to small-scale borrowers.

Expansion of export potential for small-scale industrial products would also help to overcome some problems related to low demand and small volume of production. India and Nepal provide market profiles for several export industries, assist entrepreneurs to conduct market studies, and provide trend and forecasting analyses on market demand for some exportable products.[70] Those countries that have achieved some progress in promoting small-scale industrial exports have used a variety of assistance methods. Korea and Japan have comprehensive programs that include export advance incentives, loans against export usance bills, and long-term export financing. India and Japan have drastically simplified export and customs procedures for some types of small enterprises. They have established advisory services on marketing exports and give transport and shipping concessions. Through marketing promotion schemes information about local products is disseminated abroad, exhibitions and trade fairs are sponsored by the government in foreign countries, and central government agencies coordinate export production activities for manufacturers.[71]

Korea provides tax concessions and exemptions from customs duties, offers inspection and quality-control services and technical advice to manufacturers, and builds industrial estates for small enterprises capable of producing exportable goods.[72] Similar programs are needed to expand the production of goods for the domestic market. The Japanese government has set up a large number of open laboratories, for instance, to test, analyze, and inspect raw materials for small industrialists to help them improve the quality of their goods as well as raise their productivity. Under the "traveling technical guidance" programs, university professors and technicians from private firms tour smaller factories throughout the country identifying and analyzing operational problems, offering recommendations for improvement, and holding guidance classes in various cities. Small enterprises are given access to the services of research and experimental institutes free of charge. Modernization promotion councils at the national and prefectual levels help small enterprises to expand operations and modernize their plants.[73]

Expanding the capacity of small- and medium-scale industries to strengthen the economic base and increase employment opportunities in secondary cities of developing countries depends on an extensive and integrated promotional and assistance program by national and local governments.

Enhancing Local Government Capabilities

Coping with all of the other problems of urban development requires local governments that can plan and manage public services and finance development projects that will strengthen the economics of secondary cities. Yet, in most developing countries, local governments are weak and are being squeezed from two directions. Population growth from migration and natural increase, rising expectations, and growing deficiencies in services and employment are placing greater demands on limited local revenues at a time when the economies of these cities are still weak and their governments still depend on the national government for most of their funds. Yet, the pressures on national governments have also been increasing because of slow rates of economic growth and greater demands for investment. National governments, therefore, have little flexibility to meet the needs of secondary cities. This dilemma is noted explicitly in a recent study of intermediate cities in the Philippines, which points out that "the demands being placed on the resources of the national government are such that they are increasingly precluding the possibility of [it] taking any lasting positive actions to deal with the crucial problems of poverty in the chartered cities. The responsibility and initiative in dealing with these problems are going to fall increasingly on the shoulders of the chartered city governments."[74]

Yet, local administrative units in the Philippines, as in most other developing nations are creatures of the central government. They depend on it for authority, financial resources, and technical expertise. They have little capacity to deal with the problems that accompany growth and diversification or to take the initiative in promoting development. Typically, the ability of local governments to plan, finance, and carry out programs is severely constrained, as is their ability to stimulate, guide, or support urban development, by the following factors:

(1) They lack local sources of revenue to deal with urban problems and to meet local needs and are dependent on transfers of revenue from the central government both to undertake development projects and to finance their routine operations.

(2) Many city governments are understaffed and have few competent planners and managers. Local administrative and technical personnel are poorly paid, morale and productivity is low, and political and bureaucratic constraints discourage them from taking the initiative in solving local problems.

(3) Few city governments have control over the services and facilities needed to strengthen their economies; such authority is often retained by national ministries or fragmented among local, provincial, state, and national government agencies.

(4) Severe restraints are often placed by the national government on local governments' ability to raise revenue, either through local taxes or by borrowing, as well as on their discretion in making expenditures from local budgets.

(5) City officials not only lack the technical or managerial skills to cope with complex economic, social, physical, and administrative problems, but they usually lack appropriate equipment, supplies, and facilities to provide the basic services needed to stimulate growth and to guide or manage development.

(6) There are few, if any, mechanisms through which residents of most secondary cities can participate in development decisions that affect their neighborhoods and districts and through which problems of secondary city growth can be solved.

Paradoxically, in many countries the government structure is highly centralized, but control over local functions and activities is fragmented widely among government agencies. In the Philippines, for example, mayors have the authority to appoint only a few low-level administrators, and all of the more important city government officials—treasurers, assessors, engineers, and fiscal and health officers—are appointed by national ministries, usually without consultation with the mayors.[75] In Thailand, responsibility for roads, traffic management, electricity, telephones, middle and high school education, hospitals, police protection, public housing, and other essential urban services are held by national ministries, agencies, and boards. Municipal governments have responsibility only for land use planning, fire protection, markets and parks, primary education, and refuse disposal; and they even must depend on the central government for revenues to carry out these tasks.[76] Graham notes in his study of local government in Latin America that "the impact of centralization and reconcentration of bureaucratic power in the nation's capital, no matter how dispersed its actual application may be within individual agencies and ministerial divisions, is such that few services can be provided at the local level that are not the function of some field organ of a central government ministry or autonomous agency."[77] In Brazil, the authority to provide services to secondary cities is fragmented among public agencies and corporations, local governments, and central government agencies. World Bank analysts point out that there are great differences in the capacity, and interest, of these organizations to provide services to cities and, thus, the government's development programs have had limited success in coping with urban problems.[78]

This organizational overlapping and fragmentation is not necessarily the most critical problem. Indeed, it can be an advantage in channeling resources to secondary cities from many sources. But the lack of an institution specifically responsible for implementing secondary city development programs leaves a vacuum within the national government. Unless one agency has primary responsibility for implementing policy, it is unlikely that secondary cities will have a strong advocate in the central government to press their demands and obtain resources needed for their development. Moreover, without stronger authority to deal with urban problems within the governments of secondary cities there is little chance that they can make effective plans for guiding the pace and direction of development or integrate their plans with national and local budgets. Atmodirono and Osborn observe the great difficulties governments in Indonesia's secondary cities have in "pursuading each service administration with a larger compass of responsibility (like the national electricity, education, health and telecommunications bodies) to provide what the city thinks it needs—in competition with all other localities in Indonesia."[79]

If national urbanization policies are to be carried out effectively in secondary cities, it seems essential to have a national agency at cabinet level, or a semiautonomous urban development authority, to coordinate and direct central government programs for secondary city development. Such an agency should have the power to raise revenues—through bonds, taxation, or special allocations from the treasury—to construct infrastructure, utilities, roads, and other capital facilities, and to provide essential services in secondary cities, until their revenue bases can be expanded.

Moreover, as cities grow it becomes more difficult to govern them effectively from the national capital. Greater authority to make and carry out decisions concerning development is needed within local governments. If devolution of authority and responsibility is not politically feasible, then prefectual arrangements, wherein a single chief executive is appointed by the central government to coordinate activities within secondary cities, could help to integrate the activities of government agencies.

The most serious and frequently reported difficulty that city governments face in meeting the needs and challenges of urban growth is the severe shortage of local financial resources. In most developing countries, national governments not only provide little authority for municipal and local governments to raise revenue, but place tight restrictions on their ability to do so. Thus local governments depend on transfers and grants from the national treasury. The World Bank found in studies of secondary cities in Thailand that "the local authorities' powers to raise both tax and non-tax revenues are limited and the criteria according to which shared revenues are distributed among government units are inconsistent."[80] Small secondary cities in Mexico, such as Oaxaca, are "forced to depend almost completely on allocations from the center."[81] Their success in obtaining development funds is sporadic and often they can only acquire money for projects that are visible and cosmetic, such as improving the appearance of the city hall or building a sports stadium. In Egypt, urban governorates in secondary cities such as Port Said, Ismalia, and Suez receive from 79 to 87 percent of the money for local expenditures from the central government, and all local governments receive 75 percent or more of their budgets from national grants.[82]

Local governments in many developing countries are allowed to derive revenues only from sources that are likely to yield the least amounts of money, while more lucrative sources are retained or taken over by the central government. In Thailand, for example, municipal governments depend on local revenues from rent taxes on commercial property, a land development tax—from which owner-occupied residential and agricultural land, most government land, land used for religious purposes, and commercial buildings subject to rent taxes are all exempt—slaughter and signboard taxes, rental income from municipal property, and fees and license charges. These tend to be either very restricted revenue sources or taxes that are difficult to collect. All other sources of revenues are controlled and proceeds are distributed by national agencies. Local governments lack authority even to collect local property taxes. Thus in cities such as Chiangmai, municipal development expenditures are extremely low, averaging less than the equivalent of $75,000 a year.[83] World Bank analysts argue that "the revenues of the municipalities and other local governments are inadequate for effective performance of their legally mandated responsibilities."[84] Similarly, in Indonesia, observers note that the central government "reserved to itself and secondarily the provincial governments the most lucrative taxes."[85] The cities obtain the bulk of their limited revenues from poorly administered levies on buildings, lands, hotel and restaurant receipts, and radios. They depend on central government transfers for most of their operating expenditures and development investments. Most of the services provided in large cities have little or no influence on their decisions.

Among the actions needed to improve and strengthen local financial capacity are:

(1) expanding taxing and revenue-raising authority for the larger secondary cities, allowing them to raise taxes from a wider variety of local sources and to use a greater number of taxing instruments;

(2) creating special funds for urban development that can be used to finance costly capital and infrastructure investment in secondary cities and that can be replenished from national revenue sources such as customs, excise, or import taxes that are set aside from line agency budgets;

(3) providing statutory payments to local governments from fixed percentages of recurrent revenues of central government agencies or state or provincial government budgets as grants, thus giving city administrations more flexibility to meet local needs and demands;

(4) providing technical assistance and training to local officials in improving tax administration and collection procedures and increasing revenues from existing sources;

(5) allowing city governments to draw loans from national development banks or credit authorities to provide services and

facilities for which users can be charged and that generate revenue that can be used to repay the loans;

(6) allowing more flexibility for local governments to spend their revenues to meet local needs and demands, without overly constraining rules and regulations;

(7) granting to or expanding property taxing powers for city governments and improving their capacity to do tax mapping, keep and maintain records, assess property more accurately and fairly, and collect revenues more effectively;

(8) standardizing and improving intergovernmental transfers and allocating funds from the national government to secondary cities more effectively, so that revenues reach them in a timely manner;

(9) allowing or expanding the authority of city governments to engage in public enterprise for activities not offered by the private sector, and relaxing restrictions or ceilings on the profits of such enterprises; and

(10) allowing or expanding the authority of city governments to levy preferential land use and business taxes to stimulate desired economic activities and to guide the physical patterns of development within intermediate cities.

Finally, most secondary city governments lack the skilled personnel to plan and manage development activities, especially those improving the living conditions and employment opportunities of the poor. Atmodirono and Osborn found in their studies of Indonesia that the shortage of trained technical and administrative staff was at least as serious a problem for secondary city governments as the scarcity of revenue. It inhibited them from extending services and facilities and delayed or obstructed new development projects.[86] A similar problem was noted in nearly all of the case studies of secondary cities reviewed here. Programs are needed to upgrade the planning, management, and organizational capacity of secondary city governments and to establish procedures through which lower-income families can participate in the planning and administration of self-help programs in their neighborhoods. Also needed are programs to strengthen secondary cities' administrative and organizational structures for planning and coordinating public services and facilities to conduct surveys to identify appropriate client groups and their needs to create a planning and management process that brings together city officials, national government ministry staff, and affected citizens to shape the physical, social, and economic development of the city, and to train city officials and staff in managerial and technical skills required for planning and development.[87]

The Need for Further Research and Development

Because much of the research on urbanization in developing countries in the past has been focused on macroeconomic and demographic issues and on the problems of primate cities and the largest metropolitan areas, relatively little is still known about the economic, social, and physical factors affecting the growth and development of secondary cities. This review of the literature and experience with secondary city development raises even more questions. At best, this and other studies only provide a starting point for the extensive research and hypothesis testing that remains to be done before international development organizations and national governments can begin to plan and implement urban development policies more effectively.

DIRECTIONS FOR COMPARATIVE RESEARCH

The list of issues that need further investigation is long. The East-West Center's Population Institute has identified topics that range from the implications of national and international development strategy, of national economic and social policies, and of the characteristics of labor markets and population mobility processes to the effects of resource accumulation and circulation processes and of institutional and political structures on the growth and development of secondary cities. High-priority topics include:[88]

(1) comparative research on national and international policies affecting secondary cities, including the nature of urbanization

and transformation from rural to urban societies in different
countries, studies of international emigration and the flow of
remittances, and analyses of the leakage of resources due to
different patterns of industrialization and investment;

(2) citywide or regional studies of the effects of different agricultural or industrial development policies on regional economic
growth and the development of systems of secondary cities; the
effects of decentralization and resource frontier development on
intermediate cities; the structure and processes of informal and
formal sector labor markets and circular migration in secondary
cities; resource flows through regional urban centers and their
retention rates; and the roles of elites in urban and regional
development;

(3) sectoral studies of the impact of export industrialization, multinational banking, and agrobusiness on secondary city economies, the impact of agricultural pricing policies on food production and distribution in urbanizing regions, the impact of
transportation on rural-urban and regional labor mobility, and
the impact of public expenditure patterns on secondary city
development; and

(4) household studies of the impact of inflation on family expenditures and conditions of self-sufficiency in secondary cities, the
incidence of different tax schemes on lower-income households,
social and spatial mobility of households at different income
levels living in secondary cities, and patterns of local participation and political leadership.

To these topics, Richardson has added the need for more
effective analyses of the economic and employment structures of cities of different sizes in developing countries, of
threshold levels of service costs, and of threshold sizes of
cities performing different combinations of urban functions.
He argues that little research has been done on the policies
that are likely to lead to polarization reversal, and that more
research is needed in those countries, such as Brazil and
Korea, where it seems to be taking place. Moreover, he
suggests that an important research priority is the analysis of
"what types of city size distribution offer the best compro-
mise between the hierarchy needed for the production and
distribution of goods and services and that required for the

transmission and diffusion of growth impulses and innovation over national space."[89] More research is also needed on agglomeration economies in cities of different sizes, the impact of transportation linkages on urban and regional development, and the effectiveness of different location and infrastructure investment subsidies in cities of different sizes.[90]

This study also raises a number of issues about which much more needs to be known:

(1) the relationships among population size, interurban distances, market areas, and the capacity of secondary cities to perform particular economic and social functions;

(2) the role of different patterns of national investment allocation on stimulating or retarding secondary city growth;

(3) the patterns of change in comparative advantage, specialization, and functional complexity in secondary cities as nations become more urbanized and industrialized;

(4) the interaction of economic, political, social, and physical factors in the dynamics of urban growth in countries with different national development strategies;

(5) the importance of different combinations of factors in the developmental or exploitational orientation of urban economies under different sets of political and economic conditions, and the impact of local and national policies on strengthening or changing those orientations;

(6) the roles of local leaders in promoting the growth of secondary cities as catalysts for regional development;

(7) the effects of different combinations of linkages and interactions between secondary cities and smaller towns and villages in their regions on growth and development of rural areas;

(8) the degree of linkage or "closure" needed to promote equitable development of smaller cities and towns in rural hinterlands of secondary cities;

(9) ways in which the commercial and industrial sectors of secondary cities can be made more labor absorbing and in which demand for local goods and services can be expanded in cities with high levels of poverty; and

(10) the effects of national policies for limiting or redirecting the growth of primate cities and major metropolitan areas on the growth and development of secondary cities.

Undoubtedly these lists could be expanded even further, but they clearly indicate the magnitude of the need for systematic and comparative analyses of secondary cities in developing countries. The pressures of urbanization in developing nations, however, are unlikely to abate while researchers gather more information. The most effective research will probably be applied in combination with pilot and demonstration projects that test alternative strategies and policies for changing the urbanization pattern in developing nations.

APPLIED ANALYSES AND PILOT PROJECTS

International agencies and governments in developing countries can play an important role in increasing the base of knowledge and information about secondary cities through:

(1) Comparative research on the socioeconomic and physical characteristics and on the dynamics of growth of secondary cities. Such information is needed to formulate more relevant and appropriate national strategies for urban development.
(2) Longitudinal studies of changes in secondary cities as national and political conditions change. These studies can provide better information by which policies can be formulated to reinforce or alter changing conditions.
(3) Pilot and demonstration projects that test combinations of policies and programs for strengthening the functions and stimulating the growth of secondary cities in different regions of a country.
(4) Longitudinal studies of secondary cities in different regions of developing countries to provide better information about regional variations in the dynamics of development.

Finally, it should be noted that international assistance organizations and governments in developing nations can do a great deal to strengthen the intermediate level of the urban system, simply by reorganizing many of their existing financial and technical assistance programs and by refocusing them on secondary cities. These bodies now engage in many development activities that, directly or indirectly, affect the growth of secondary cities and that could be used to strengthen the economies of such cities if the activities were more integrated and deliberately designed to develop secondary urban centers. Similar programs for smaller cities and large market towns in rural areas might stimulate their growth and diversification as well, thereby increasing the number and geographical distribution of secondary cities and creating a stronger network through which the benefits of urbanization might be spread more equitably.

NOTES

1. World Bank, *Brazil—Medium Sized Cities Project* (Washington, DC: Author, 1979), p. 1.

2. Ibid.

3. Arthur D. Murphy and Henry A. Selby, "Poverty and the Domestic Life Cycle in an Intermediate City of Mexico" (Paper prepared for the Workshop on Intermediate Cities, East-West Center Population Institute, Honolulu, 1980), pp. 5-7.

4. U.S. Agency for International Development, *Rural Service Center Project Paper* (Manila: Author, 1977), p. 12A.

5. Robert A. Hackenberg, *Fallout from the Poverty Explosion: Economic and Demographic Trends in Davao City, 1972-1974* (Davao: Davao Action Information Center, 1976), p. 3.

6. See I. Adelman, C. Morris, and S. Robinson, "Policies for Equitable Growth," *World Development* 4, no. 7 (1976): 561-582; Irma Adelman and Cynthia T. Morris, "Growth and Impoverishment in the Middle of the Nineteenth Century," ibid 6, no. 2 (1978): 245-273.

7. James Osborn, *Area Development Policy and the Middle City Under the Indonesian Repelita as Compared to the Malaysian Case: A Preliminary Analysis* (Santa Barbara, CA: Center for the Study of Democratic Institutions, 1974), p. 50.

8. World Bank, *Thailand: Urban Sector Review,* Background Working Paper 7 (Washington, DC: Author, 1978), p. 26.

9. Akin L. Mabogunje, "The Urban Situation in Nigeria," in *Patterns of Urbanization: Comparative Country Studies,* ed. S. Goldstein and D. F. Sly

(Liege, Belgium: International Union for Statistical Study of Population, 1977), pp. 614-615.

10. Walter F. Weiker, *Decentralizing Government in Modernizing Nations: Growth Center Potential of Turkish Provincial Cities* (Beverly Hills, CA: Sage, 1972), p. 39.

11. U.S. Agency for International Development, *Rural Service Center Project Paper.*

12. Hackenberg, *Fallout from the Poverty Explosion,* p. 148.

13. Ibid., p. iv.

14. World Bank, *Brazil–Medium Sized Cities Project,* p. 2.

15. PADCO, Inc., *Guidelines for Formulating Projects to Benefit the Urban Poor in the Developing Countries,* vol. 2 (Washington, DC: USAID, 1976), pp. III-5ff.

16. Ibid., pp. II-22-II-39.

17. Ibid.

18. Ibid., pp. II-6, II-7.

19. See R. L. Meier, S. Berman, T. Campbell, and C. Fitzgerald, *The Urban Ecosystem and Resource Conserving Urbanism in Third World Cities* (Washington, DC: USAID Office of Urban Development, 1981).

20. Mabogunje, "Urban Situation in Nigeria," pp. 611-612.

21. U.S. Agency for International Development, *Philippines Shelter Sector Assessment,* vol. 1 (Washington, DC: Author, 1978), pp. 3-4.

22. Joseph H. Chung, *Housing and Residential Land in Korea: An Overall Evaluation* (Seoul: Korea Research Institute on Human Settlements, 1980), pp. 26, 32, 37.

23. Lachlin Currie, *Accelerating Development: The Necessity and the Means* (New York: McGraw-Hill, 1966).

24. Janet L. Abu-Lughod, "Developments in North African Urbanism: The Process of Decolonization," in *Urbanization and Counterurbanization,* ed. B.J.L. Berry (Beverly Hills, CA: Sage, 1976), pp. 191-212.

25. Ibid.; see especially pp. 202-210.

26. Herbert R. Barringer, "Conclusions," in *A city in transition: Urbanization in Taegu, Korea,* ed. Man-Gap Lee and Herbert R. Barringer (Seoul: Hollym, 1971), pp. 565-578.

27. Jay Weinstein and V. K. Pillai, "Ahmedabad: An Ecological Perspective," *Third World Planning Review* 1, no. 2 (1979): 209.

28. Cited in Harold Lubell, *Urban Development Policies and Programs,* Working Paper for Discussion, Economic Development Division (Washington, DC: USAID, 1979), p. 19.

29. Chung, *Housing and Residential Land,* p. 46.

30. Weiker, *Decentralizing Government,* p. 56.

31. See Jorge E. Hardoy, "Urbanization Policies and Urban Reform in Latin America," in *Latin American Urban Research,* vol. 2, ed. G. Geisse and J. E. Hardoy (Beverly Hills, CA: Sage, 1972), pp. 19-44.

32. See Clifton W. Pannell, *T'ai-Chung, T'ai-Wan: Structure and Function,* Research Paper 144 (Chicago: Department of Geography, University of Chicago, 1973).

280 SECONDARY CITIES IN DEVELOPING COUNTRIES

33. Chakrit Noranitipadungkarn and A. Clarke Hagensick, *Modernizing Chiengmai: A Study of Community Elites in Urban Development* (Bangkok: National Institute of Development Administration, 1973), p. 14.

34. See D.C.I. Okpala, "Housing Standards: A Constraint on Urban Housing Production in Nigeria," *Ekistics* 45, no. 270 (1978): 249-257.

35. Weiker, *Decentralizing Government*, p. 61.

36. Malcolm D. Rivkin, *Land Use and the Intermediate Size City in Developing Countries* (New York: Praeger, 1976).

37. Malcolm D. Rivkin, "Some Perspectives on Urban Land Use Regulation and Control," in *Urban Land Policy Issues and Opportunities,* World Bank Staff Working Paper 283, World Bank (Washington, DC: World Bank, 1978), pp. 85-126.

38. Mabogunje, "Urban Situation in Nigeria," p. 615.

39. See Bertrand Renaud, *National Urbanization Policies in Developing Countries,* World Bank Staff Working Paper 347 (Washington, DC: World Bank, 1979).

40. Johannes F. Linn, *Policies for Efficient and Equitable Growth of Cities in Developing Countries,* World Bank Staff Working Paper 342 (Washington, DC: World Bank, 1979), p. 171.

41. Ibid., pp. 169-172.

42. Frederick Temple et al., *The Development of Regional cities in Thailand* (Washington, DC: World Bank, 1980), p. 31.

43. Renaud, *National Urbanization Policies,* p. 81.

44. The literature is reviewed and some tests of the relationship are made by Alan Gilbert, "Transport Improvement and Rural Outmigration in Colombia," *Economic Development and Cultural Change* 29 (April 1981): 613-629.

45. Rivkin, "Some Perspectives on Land Use Regulation," p. 97.

46. International Labor Organization, *Towards Full Employment: A Programme for Colombia* (Geneva: Author, 1970), pp. 357-360.

47. Hackenberg, *Fallout from the Poverty Explosion,* p. 8.

48. Ibid.

49. Ray Bromley, "Organization, Regulation and Exploitation in the So-Called 'Urban Informal Sector': The Street Traders of Cali, Colombia," *World Development* 6, nos. 9, 10 (1978): 1165.

50. Roy Bahl et al., *Strengthening the Fiscal Performance of Philippine Local Governments* (Syracuse, NY: Syracuse University Local Revenue Administration Project, 1981), p. ES33.

51. See World Bank, *Employment and Development of Small Enterprises Sector Policy Paper* (Washington, DC: Author, 1978).

52. Hackenberg, *Fallout from the Poverty Explosion,* pp. 202-203.

53. PADCO, Inc., *Guidelines for Formulating Projects,* p. I-9.

54. This section draws heavily on Dennis A. Rondinelli, "Small Industries in Rural Development: Assessment and Perspective," *Productivity* 19, no. 4 (1979): 457-480.

55. Hackenberg, *Fallout from the Poverty Explosion,* p. 15.

56. Bryan Roberts, "The Social History of a Provincial Town: Huancayo, 1890-1972," in *Social and Economic Change in Modern Peru,* ed. R. Miller, C. T.

Smith, and J. Fisher (Liverpool: Center for Latin American Studies, University of Liverpool, 1976), pp. 136-197.

57. More detailed discussions of these points can be found in Rondinelli, "Small Industries in Rural Development."

58. See Arthur Gibb, Jr., "Local Industries: Nonagricultural Production and Employment in Agricultural Regions," mimeographed (Washington, DC: USAID, 1974).

59. Frank C. Fairchild and Hiromitsu Kaneda, "Links to the Green Revolution: A Study of Small Scale, Agriculturally Related Industry in the Pakistan Punjab," *Economic Development and Cultural Change* 23 (January 1975): 249-275.

60. See P. M. Mathai, "Rural Industrialization and the Maximization of Employment Opportunities in India," *Small Industry Bulletin for Asia and the Far East* 9 (1974): 59-63.

61. See Asian Productivity Organization, *Project Feasibility Study Training Courses: Stages I and II* (Tokyo: Author, 1977).

62. Medium Industry Bank, "Small Industries Policies: Republic of Korea," *Small Industry Bulletin for Asia and the Far East* 8 (1971): 84-86.

63. World Bank, *Thailand: Urban Sector Review,* p. 48.

64. Ibid., p. 41.

65. Russell J. Cheetham and Edward K. Hawkins, *The Philippines: Priorities and Prospects for Development* (Washington, DC: World Bank, 1976), p. 240.

66. See David Kochan et al., *Financing the Development of Small Scale Industries,* World Bank Staff Working Paper 191 (Washington, DC: World Bank, 1974), pp. 11-41.

67. See United Nations Economic Commission for Asia and the Far East, "Seminar on Financing Small Scale Industry in Asia and the Far East: Conclusions and Summary," *Small Industry Bulletin for Asia and the Far East* 9 (1972).

68. See M. A. Oommen, "A Report on Small-Scale Industry Development Based on a Study Tour of Japan, the Philippines and Thailand," *Small Industry Bulletin for Asia and the Far East* 8 (1971): 106-119.

69. See H. Ansari, "A Brief Discussion on the Merging of Small Industries in Iran," *Small Industry Bulletin for Asia and the Far East* 11 (1973): 2-5.

70. See S. Rana, "The Place of Marketing in Programmes for Promotion, Modernization and Development of Small-Scale Industries," *Small Industry Bulletin for Asia and the Far East* 13 (1976): 48-53.

71. See P. Nariasiah, "Export Promotion of Small Industry Products," *Small Industry Bulletin for Asia and the Far East* 7 (1970) 33-44.

72. Kok Chong Yu, "Measures for Promoting Exports for Small Industry Products in the Republic of Korea," *Small Industry Bulletin for Asia and the Far East* 7 (1970): 26-30.

73. Oommen, "Report on Small-Scale Industry Development," pp. 115-117.

74. U.S. Agency for International Development, *Rural Service Center Project Paper,* p. 34.

75. Ibid., pp. 51-54.

76. World Bank, *Thailand: Urban Sector Review,* pp. 65-68.

77. Lawrence S. Graham, "Latin America," in *International Handbook on Local Government Reorganization,* ed. Donald C. Rowat (Westport, CT: Greenwood, 1980), p. 489.

78. World Bank, *Brazil–Medium Sized Cities Project,* pp. 3-4.

79. Abukasan Atmodirono and James Osborn, *Services and Development in Five Indonesian Middle Cities* (Bandung: Institute of Technology, Center of Regional and Urban Studies, 1974), p. 44.

80. World Bank, *Thailand: Urban Sector Review,* p. 68.

81. Meier et al., *The Urban Ecosystem,* p. 71.

82. See James B. Mayfield, *The Budgetary System in the Arab Republic of Egypt: Its Role in Local Government Development* (Washington, DC: USAID, 1977).

83. World Bank, *Thailand: Urban Sector Review,* p. v.

84. Ibid., p. 68.

85. Atmodirono and Osborn, *Services and Development,* p. 66.

86. Ibid.

87. U.S. Agency for International Development, *Rural Service Center Project Paper.*

88. J. T. Fawcett, R. J. Fuchs, R. Hackenberg, K. Salih, and P. C. Smith, *Intermediate Cities in Asia Meeting: Summary Report* (Honolulu: East-West Center Population Institute, 1980), pp. 11-14.

89. Harry W. Richardson, *City Size and National Spatial Strategies in Developing Countries,* World Bank Staff Working Paper 252 (Washington, DC: World Bank, 1977), pp. 71-73.

90. Ibid., pp. 72-75.

INDEX

ABOUT THE AUTHOR

DENNIS A. RONDINELLI is Professor of Development Planning at the Maxwell School of Citizenship and Public Affairs, Syracuse University. He received his Ph.D. from Cornell University. He has taught at the University of Wisconsin and Vanderbilt University, and has served as senior fellow at the East-West Center in Honolulu, Hawaii, and at the Graduate School of International Management at Chung-Ang University in Seoul, Korea. Dr. Rondinelli has served as a consultant to development projects in Asia, Africa, and Latin America, and has been an adviser to the U.S. Agency for International Development, the United Nations Centre for Regional Development in Nagoya, Japan, and the World Bank. He has published over 60 articles on various aspects of international development policy and urban and regional development planning, and is author, coauthor, or editor of 7 books on development policy, planning, and administration.